Water, Rock, & Time

The Geologic Story of Zion National Park

Robert L. Eves

ZION NATL PARK FOREVER PROJECT

ISBN 978-0-915630-42-4
Second Edition
10 9 8 7 6 5 4 3 2 1
First edition published 2005

Project managed by **Lyman Hafen**

Book design by **Sandy Bell Design**

Illustrations by **Rick Wheeler**

Printed in Hong Kong
through QuinnEssentials

FRONT COVER:
The erosive process of the Virgin River carved the deep Narrows of Zion Canyon over the last two million years.

BACK COVER:
Climbers enjoying the spectacular sandstone walls of Zion.

ENDPAPERS:
Through the ages the swift and persistent Virgin River ground away at its bed of sandstone and shale in Zion Canyon.

PAGE 1:
Near the Subway in the Left Fork of North Creek the stream runs in a joint creating the beginning of a joint canyon.

PAGES 2–3:
Rushing waters of the Virgin River wind through fallen boulders as the process of canyon formation continues.

THIS PAGE:
A hiker rappels into the heart of Englestead Canyon, a branch canyon feeding The Narrows.

CONTENTS

"In all that wondrous expanse of magnificent precipices we hear no sound save our own voices and the whisper of the wind that comes and goes, breathing with the round of the centuries."

—FREDERICK DELLENBAUGH, *Scribners' Magazine*, January 1904

The labyrinth of canyons known as Zion National Park are seen here in morning light via an aerial view from above the east entrance.

Foreword

A Work in Progress

N OT LONG AGO, on a magnificent summer day, I sat on a rock near the top of Blowhard Peak on Cedar Mountain and gazed southwestward. Below me, an expansive landscape of pine-studded terraces, winding ravines and endless humps and hollows stretched into the distance and broke into labyrinths of canyons barely visible in the hot haze. My eyes strained to take it all in and finally, in the deepest distance, I discerned the familiar shape of Zion's West Temple. There, at the foot of that massive sandstone tower my office sits. I often find myself gazing through its window, wondering at the unfathomable energy nature generated to form that stunning skyline. Sitting atop the mountain overlooking much of the Virgin River's drainage in one gulp of sight, I could only begin to comprehend how water, wind, and time have patiently created the place we now call Zion National Park.

Over the past two million years the Virgin River has been the key player in Zion Canyon's formation. The Virgin originates at 9,000 feet (2,743 m) above sea level on the Markagunt Plateau. From its alpine origins among aspen and spruce along the southern flank of Cedar Mountain, the North Fork of the Virgin gathers the modest issue of a few key springs and the winter runoff from a thousand draws and ravines, consolidating trickles and brooklets into a channel that tumbles down through the strata of time, moving ever downward on its perennial push to the sea. In the process, this deceptively simple river sculpted one of the most glorious canyons on Earth. The river courses some 160 miles (257 km) before entering what was once the Colorado River at Lake Mead. The Virgin River's average rate of fall is 48 feet per mile (23.4 m per km), but it steepens to 76 feet per mile (37.1 m per km) through Zion Canyon.

Through the ages the swift and persistent stream, averaging 100 cubic feet (2.8 m³) per second, has ground at its bed of sandstone and shale, pack-

In Zion Canyon you are engulfed in time.

ing the loosened stone away at the rate of 120 cubic yards (92 m³) a day. During flood stage the process goes wild as torrents of ten times normal can remove two thousand times as much silt, sand and gravel. As one early naturalist of Zion remarked, "This is one Virgin that gets away with a lot of dirt."

In Zion Canyon you are engulfed in time. Time below you, time above you, time laid down beside you in the minuscule grains of sand that form the rock your shoulder brushes.

The great anthropologist Richard Leakey once compared the age of the Earth to a thousand-page book. If each page represented four-and-a-half million years, he surmised, the age of the dinosaurs would not begin until page 728, and all recorded history would fit comfortably on the last line of the last page. Here in Zion Canyon those pages lie atop one another in layers of sedimentary rock. The higher we climb—each 12 to 15 inches (30 to 38 cm) represents about a thousand years of erosion—the closer we come to that last page of which human beings are such a small part.

Several years ago my wife and I set out on the switchback trail to Observation Point. As we began the ascent we imagined how we were working our way up through time, drawing nearer to the clouds, and nearer to the youngest rock layers with each millennial step. We stopped at the corner of a switchback and sat for a moment. Enfolded in the brisk air of morning shadow we looked off to the west where Angels Landing rose gracefully in the early light. The West Rim stood in soft relief, its walls, ledges and ridges plunging and

ABOVE: The walls of Zion Canyon in this southward view from Angels Landing tower more than 2,000 feet (610 m) above the Virgin River.

OPPOSITE LEFT: The sandstone walls of Zion Canyon are evidence of two million years of erosion.

INSET OPPOSITE: A lone ponderosa pine has miraculously found root in a Navajo Sandstone pinnacle near Checkerboard Mesa.

towering in such majestic shapes it made my heart race. I marveled at how water, wind, and time have created this canyon. These were, and continue to be, the sculptor's tools.

At my feet appeared a whiptail lizard. It stopped, facing me, and froze erect and alert. It stood there in still life, not moving a flicker. I watched for a long, long time in the context of my jumbled and easily distracted thought processes. Maybe a minute. But what is a minute to a reptile whose ancient eyes and cold blood are but a generation removed from the Jurassic? What is a minute in this deep, deep canyon where a millennium is but a heart beat?

It was only a minute the lizard stood there—and an eternity.

In 1995, I sat in a chamber of commerce meeting and watched a program by Don Falvey, then superintendent of Zion National Park. He showed photos of an April 1995 landslide in Zion Canyon. The slide briefly dammed the Virgin River and, in an effort to find a new course, the river washed out 200 yards (183 m) of the main canyon road. Mr. Falvey related how a similar landslide had occurred in 1941, and another in 1923. At the end of his presentation, as local business leaders finished their ice cream sundaes and settled back in their chairs, Mr. Falvey took questions. The first man who raised his hand addressed the superintendent with grave concern. "Looks to me like we've got a trend starting here," he said. "We had a major slide in the early

ABOVE: A sandstone pocket filled with water from a summer storm on the park's east side.

BELOW: A winter dusting of snow on Zion's cliffs and canyon floor.

part of the century, one at mid-century, and now here we are at the end of the century with another one. What are you going to do about this?"

I don't know how many of those present caught the irony in the man's question. A trend? Yes, most certainly a trend. A trend that started about two million years ago. A trend that will continue as long as the Earth circles the sun and gravity exerts its pull, as long as clouds form rain, and rain falls, and water flows the course of least resistance. As for what he was going to do about it, Mr. Falvey replied that there was nothing in his job description requiring him to stand in the way of the ongoing process of canyon creation. In fact, it was his mission as superintendent to see that the process continued unimpaired.

Of course it would be possible for man to intervene. We can literally move mountains if we make up our minds to. We can achieve in days what water, wind, and time would spend a thousand years doing. Yet as we putter around with our bulldozers, backhoes, and belly dumpers, altering the landscape to fit our whims, nature continues her slow, persistent work.

One summer afternoon our family set out for Zion National Park with plans for an afternoon hike and picnic. As we made the steep pull up the Hurricane Fault's limestone ledges, the sky darkened. A summer storm moved in. Jetting along the highway beneath Hurricane Mesa we watched the rain begin—softly at first. Then it pelted the car so furiously we couldn't hear one another talk. The kids wanted to turn around and head home, but I assured them it would be worth continuing. I was working on two assumptions. One, that summer storms in this country can vanish as quickly as they appear. And two, that there is no more magical place in the world than Zion Canyon after a summer rain.

Through the whipping windshield wipers we could see the usually serene Virgin River transform into a monster. Minute by minute the river swelled and soon raced dark brown, lashing its banks, charging like a runaway horse.

By the time we got to Rockville the rain and wind had reached gale force. Water spread down the streets in thick sheets. The slapping wipers could not keep pace with the rain and we were forced to pull over beneath a giant cottonwood and wait out the storm.

When the rain finally let up we drove on, but just beyond Rockville we had to

There is no more magical place in the world than Zion Canyon after a summer rain.

stop again. A normally dry ravine cutting through the Shinarump ridge to the north was now a raging brown waterfall. The volume of water roiling out of the rocks above was much greater than the culvert under the road. The flood crested the bridge and now quite indiscriminately slashed toward the Virgin River.

We all got out of the car and shuffled as close to the torrent as we dared. It was a magnificent and terrifying sight, but the sound made the memory all the more powerful. It was the sound of unbridled fury, water and rock clashing, an entire symphony percussion section gone wild.

We stood reverently and watched, deafened by the roar, speechless at the power manifest before us. The water, dark with sediment gathered from the plateau above, crashed and pounded and exploded toward lower ground. Into the swollen Virgin River the torrent flowed, packing part of the landscape with it, carrying on the two-million-year-old sculpting process right before our eyes.

We appreciated the spectacle for what it was—a rare and precious sight. Such opportunities to witness the sculptor's strokes within a time frame suited to our own fleeting lives are few and far between.

There was another experience. This one more powerful—and more horrifying in its instant potential to transform the landscape. In the dark morning of September 2, 1992, my wife and I awoke to the very palpable sound and the most distinct feeling that a thousand freight trains had jumped their tracks and were ploughing in our direction. The rumble came from

the southeast and rolled inexorably toward us. We shot out of sleep and tried desperately to get hold of something stable. But there was nothing stable in our world at that moment. The house shook as the trains roared through, and we felt them rumble on into the night, on to the northwest and finally into silence.

In the breathless calm that followed we reached for one another and trembled. We had just experienced an earthquake. We realized then what we had always known but never admitted—that the solid foundation we live on is not so solid. That the mountains standing so firmly around us are capable of moving.

A Richter magnitude 5.8 earthquake occurred at 4:26 A.M. Mountain Daylight Time that day. Its epicenter was about six miles southeast of St. George, Utah, ten miles southeast of our Santa Clara home. The shock was the strongest felt in southwestern Utah since 1902, and the largest in the Utah region since 1975. No deaths or serious injuries resulted from the quake, but it caused physical damage 95 miles (153 km) from its epicenter. The most visible destruction was a landslide in Springdale at the entrance to Zion National Park. The slide, measuring roughly 1,600 feet (488 m) from top to toe and about 3,600 feet (1,097 m) in width, temporarily blocked State Route 9 leading into and out of Zion. Some 18 million cubic yards (14 million m³) of rock were displaced in the slide as it moved slowly and continually for several hours after the quake. Two water tanks, several storage buildings, and three homes were destroyed. Here, the visible effect of the earthquake was poignantly graphic. But the more far-reaching effect of that early-morning earth shift was the heightened reality it evoked in thousands of people's minds—our landscape is still a work in progress.

As I sat on Blowhard Peak that day, I tried to think of myself in the context of Zion Canyon's existence. Regardless of how constant the canyon walls appeared to me on that day in this decade in this century, I knew I was actually looking at a work in progress. There was a time in ages past when the canyon did not exist, but now, as long as rain falls, winds blow and time passes, the canyon will change. The small moment of our mortal lives affords us few precious glimpses of the brush strokes that change the picture—a rock slide here, a flash flood there, an earthquake's unfathomable power. Our lives in this landscape are lived within an infinitely larger context and we can only count ourselves blessed to be surrounded by all the beauty the process of change offers.

LYMAN HAFEN
Executive Director, Zion Natural History Association

ABOVE: Dusk over Phantom
Canyon.

OPPOSITE LEFT: Bigtooth maple
leaves at Emerald Pools turn
resplendent red in autumn.

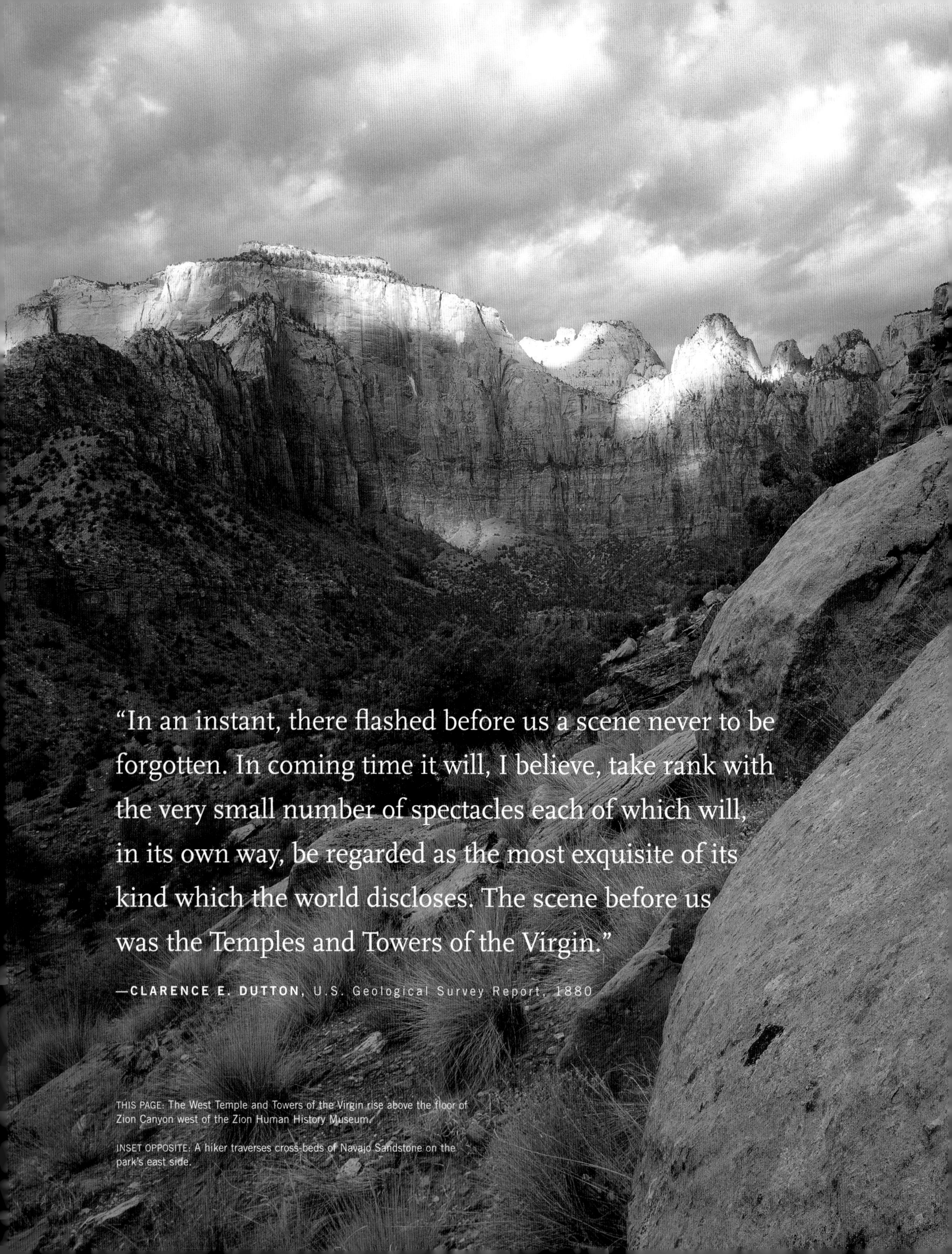

"In an instant, there flashed before us a scene never to be forgotten. In coming time it will, I believe, take rank with the very small number of spectacles each of which will, in its own way, be regarded as the most exquisite of its kind which the world discloses. The scene before us was the Temples and Towers of the Virgin."

—CLARENCE E. DUTTON, U.S. Geological Survey Report, 1880

THIS PAGE: The West Temple and Towers of the Virgin rise above the floor of Zion Canyon west of the Zion Human History Museum.

INSET OPPOSITE: A hiker traverses cross-beds of Navajo Sandstone on the park's east side.

Introduction

A Geological Overview

IN ALL THE WORLD there is no place quite like Zion National Park. "There is almost nothing to compare to it," wrote Frederick Dellenbaugh in one of the first nationally published articles on Zion Canyon. His essay appeared in *Scribners' Magazine* soon after a visit to the canyon in 1903. Responding to his first encounter with Zion's West Temple, Dellenbaugh wrote, "Niagara has the beauty of energy; the Grand Canyon, of immensity; the Yellowstone, of singularity; the Yosemite, of altitude; the ocean of power; this Great Temple, of eternity. . ."

Indeed, the ancient rock statuary of Zion National Park can beget a sense of the eternal as you contemplate the seemingly countless ages through which the canyon formed. There are many fascinating stories associated with this place, but all the stories begin with the landscape itself.

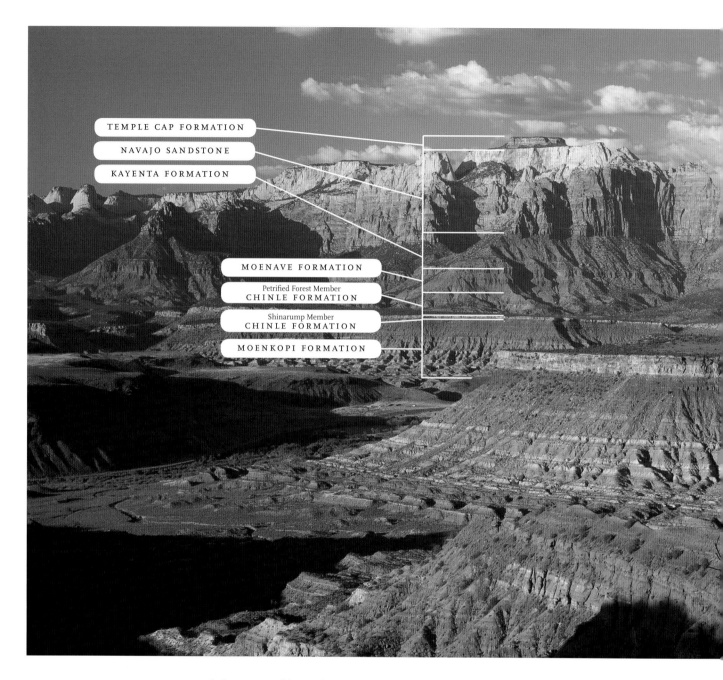

TEMPLE CAP FORMATION

NAVAJO SANDSTONE

KAYENTA FORMATION

MOENAVE FORMATION

Petrified Forest Member
CHINLE FORMATION

Shinarump Member
CHINLE FORMATION

MOENKOPI FORMATION

The awesome panorama looking northeastward from Gooseberry Mesa offers an excellent cross-section of Zion's geologic formations.

And the story of how this landscape came to appear as it does today is one of the most fascinating stories of all.

Zion National Park is probably best known for Zion Canyon, a spectacular chasm between sheer walls of Navajo Sandstone towering more than 2,000 feet (610 m) above the canyon floor. The width of Zion Canyon varies from a slim slot called The Narrows in its upper reaches, to a wider valley where the Virgin River cuts into less resistant layers of sandstone called the Kayenta and Moenave Formations. But always, the steep walls of Navajo Sandstone, capped by temples and towers, dominate the landscape.

The canyon walls are covered with blind arches and alcoves, hanging val-

leys, and colorful surface stains. The canyon floor records the evolution of the Virgin River and its interaction with the canyon walls, as well as abandoned terraces where the river ran in ancient times, and the evidence of what were once large, landslide-dammed lakes.

Many park visitors do not venture much beyond Zion Canyon, but those that do discover the canyon itself contains only a fraction of a larger geologic history. This bigger story is recorded in such places as the Kolob Canyons, the areas above the rim of Zion Canyon, and in the park's southern portion.

The story of Zion National Park is revealed in the sedimentary rocks exposed within its boundaries. Strata as old as the Early Permian Toroweap

and Kaibab Formations are visible in the Hurricane Fault zone, near the Kolob Canyons. The Early Triassic Moenkopi and Late Triassic Chinle Formations are best observed as you drive toward Springdale from Hurricane. These formations stand as mesas in the park's southern end. The dominant cliffs of Navajo Sandstone, so evident in Zion Canyon, are topped by the Temple Cap and Carmel Formations, and are best observed in the park's northern and northeastern portions. In all, nearly 7,000 feet (2,130 m) of sedimentary strata are exposed in Zion National Park, and held within these layers is the story of over 250 million years of Earth history.

From prehistory to the present, the landscape of Zion National Park has inspired wonder and awe. Early visitors could do nothing more than take it all in, but the great nineteenth century scientific expeditions into the Colorado Plateau (see map pg. 22) and Colorado River regions brought a desire to understand the area's geologic history. Many have contributed to the geologic story presented in this book. Of particular note is Herbert E. Gregory, who is credited with producing the first detailed geologic account and map (1950) of the Zion National Park region. Wayne L. Hamilton produced an even more detailed map (1978) of the area, and wrote a popular account (1984) of the park's geology. The most recent and significant summary of the park's geology was prepared by geologists from the Utah Geological Survey including Robert F. Biek, Grant C. Willis, Michael D. Hylland, and Hellmet H. Doelling. Their contribution to this presentation of the geologic story of Zion National Park is particularly significant.

Throughout its 250-plus-million-years of history, Zion National Park has evolved in unique geological ways, creating the varied and fascinating landscape that park visitors view at this particular instant in geologic time. The story is an engrossing one, a geologic history dominated by water, rock, and time.

ABOVE: Dr. Herbert E. Gregory (1869–1952) is credited with producing the first detailed geologic account and map of the Zion National Park region.

LEFT: The Virgin River packs an average of 120 cubic yards (92 m³) of loosened stone out of the canyon every day.

UPPER RIGHT AND BELOW: Dr. Gregory's compass and knapsack are part of the Zion National Park permanent collection.

[19]

THE VIRGIN RIVER is the primary agent in the formation of Zion Canyon. But the river has been aided by other erosional forces which have combined to sculpt the canyon, creating the spectacular vistas we see today. Erosion is a general term for the combined processes that constantly alter, and ultimately wear down, the surface of our planet. Erosion accompanies weathering—the biological, chemical, and physical breakdown of materials at the Earth's surface. Erosion also includes transportation of weathered debris of all sizes by gravity. This much overlooked process, which geologists call mass wasting, interacts with the river's running water to continuously move material downstream. The ultimate destinations of running water, and the weathered debris it contains, are the world's oceans. Through these combined processes Zion National Park has evolved over the past several million years.

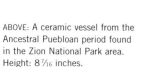

Human Exploration of Zion National Park

The first explorers of what would one day become Zion National Park were paleo-hunters who roamed the canyon, its associated tributaries, and nearby mesas. They took advantage of the sheltering alcoves and available water. Over time, these ancient peoples wandered less, and began to construct shelters. Along with the shelters came a society that not only hunted but also gathered seasonal wild fruits, vegetables, roots, and seeds.

As time passed, agriculture was introduced into the area, starting in the fertile flood-plains adjoining what would later be known as the Virgin River. The most successful crops were corn (maize), beans, squash, and other legumes. These peoples continued to gather and harvest the canyon's sea-sonal bounty. In the intervening years, Zion Canyon became home to Ancestral Puebloans (previously known as the Virgin Anasazi) and, much later, the Southern Paiutes.

Though there were undoubtedly several thousand years of native use and occupation, for some reason the canyon escaped the notice of early Spanish and American explorers. The first European reported to have viewed the canyon was Nephi Johnson, a Mormon missionary who was led there by a Paiute guide in 1858. A few years later, Isaac Behunin built a cabin in the canyon, and, along with other pioneer families, tried to farm the Virgin River floodplain. Behunin believed he had found a haven of beauty and solitude that he related to the Biblical concept of Zion, a word synonymous in Mormon culture with sanctuary. After years of farming, floods and droughts, Behunin sold his holdings in what was being called Zion Canyon and moved away in 1872.

Between 1867 and 1879, several significant geographical and geological surveys explored the American West. The U.S. Geological Survey (USGS) commissioned these early explorations which led to the discovery of, and intrigue with, the southwestern United States.

The same year Behunin moved from the canyon, Major John Wesley Powell, having finished his second voyage in the Grand Canyon, was inspired to complete the task of surveying the whole Colorado River region. His exploration party included geologists Clarence E. Dutton and Grove Karl Gilbert, photographer Jack Hillers, and artist William H. Holmes, all famous names in the history of scientific exploration in the American West. Over the next two decades this intrepid group pieced together a picture of time and the land on a giant scale. They theorized how, through periods of geologic history, the Colorado Plateau was uplifted while river systems steadily cut down through numerous layers of sedimentary rocks.

ABOVE: A ceramic vessel from the Ancestral Puebloan period found in the Zion National Park area. Height: 8 7/16 inches.

BELOW: The Crawford Family farm was located near the present-day Zion Human History Museum. The Crawfords sold their property to the National Park Service in the early 1930s.

[20]

Powell and Gilbert made a geological reconnaissance of the Zion Canyon region in 1872 and named several features. Later, Dutton directed extensive mapping of the area. His reports, along with Hillers' photographs and Holmes' drawings, first created national fascination with what would become Zion National Park. In 1903, artist Frederick S. Dellenbaugh painted scenes in the canyon and exhibited his work at the St. Louis World's Fair in 1904. He also wrote an article, published in *Scribners' Magazine,* which extolled the area's beauty inspiring others to visit the canyon.

During the summer of 1908, Leo A. Snow, a United States Deputy Surveyor, was assigned a survey in southwestern Utah which included the upper portion of Zion Canyon. Snow's crew found much of the canyon country impossible to survey, but they were struck by its incredible vistas and wild beauty. They compared Zion to the Grand Canyon. In his 1909 report to Washington, Snow stated he believed Zion Canyon ought to be designated a national park.

Snow's report, and the public interest generated by Dellenbaugh's paintings and article, led President Taft to proclaim the area Mukuntuweap National Monument on July 31, 1909. Mukuntuweap was the Paiute word Powell used for the main fork of Zion Canyon. The U.S. Congress established Zion National Park in 1919. Additional park areas were added in 1937 and 1956.

[21]

In the intervening years, the park has attracted the attention of other geologic notables. Though most famous for his work in other parts of the Colorado Plateau, in 1950, Herbert E. Gregory published *The Geology and Geography of the Zion Park Region: Utah and Arizona,* a U.S. Geological Survey professional paper. The fieldwork for this volume was completed either during Gregory's "vacation studies," (summers away from his position at the Bishop Museum in Hawaii) or as a volunteer (working mostly for the USGS for the princely sum of $1 per year). This impressive study, written by a pioneering geologist who studied Colorado Plateau geology for more than four decades, has continued to spark geological interest in Zion National Park. In some geological circles and textbooks, Zion Canyon and its features are the most discussed, most illustrated region in North America. Its uniqueness continues to inspire scientific study, the most notable and recent example of which is the work of Robert F. Biek, Grant C. Willis, and their colleagues at the Utah Geological Survey. Their 2003 publication entitled *Geology of Zion National Park, Utah,* contains the latest summary of geologic knowledge of this remarkable area.

Physiography of the Colorado Plateau

North America is divided into physiographic regions based on their common structure, geologic history, and surface features. Each of these physiographic provinces is a unique piece of North America's framework. Zion National Park is in the southwest portion of the Colorado Plateau physiographic province. Originally named the "Colorado Plateaus" by explorer John Wesley Powell, the plateau is in fact a huge basin ringed by highlands and filled with plateaus. The Colorado Plateau province covers a land area of 130,000 square miles (209,200 km²). When asked to explain what makes the Colorado Plateau unique, geographers are vague. Geologically, the Colorado Plateau is best defined by what did not happen. While the Rocky Mountains to the east, and the Basin and Range to the west and south were modified by tectonic forces, the Colorado Plateau remained structurally intact.

The Colorado Plateau is centered in the Four Corners region, where the states of Colorado, New Mexico, Arizona, and Utah meet at a single point. This region is characterized by relatively flat-lying sedimentary rocks, exposed in a series of plateaus. Home to some of the most spectacular scenery in the Southwest, the Colorado Plateau is more than a mile above sea level, rising to that height nearly as a single unit.

RIGHT: Zion National Park is found midway up the Grand Staircase, a unique unveiling of Earth's geologic history from the Grand Canyon to Cedar Breaks National Monument and Bryce Canyon National Park.

BELOW: The Colorado Plateau province covers a massive land area in Utah, Colorado, New Mexico, and Arizona.

[22]

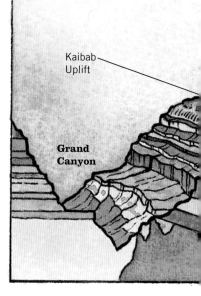

Kaibab Uplift

Grand Canyon

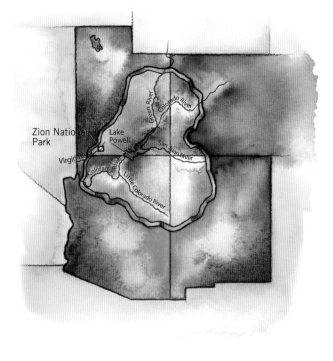

The Grand Staircase

At the Colorado Plateau's southern margin, uplift reveals the Plateau's rocky composition in a distinctive stair-like formation. Universally known as the Grand Staircase, the formations contain excellent exposures of Earth history in each step. We might imagine each step as a book, each formation as a chapter in the book, and each rock layer as a page. The books are stacked and slightly offset with the spines facing northward. Each page records, in the special language of geology, past events—shoreline migrations of a tropical marine sea, shifting sands of a vast desert, meandering flows of rivers to the sea. The books have been set aside for all to read and enjoy in the region surrounding Zion National Park.

Though its origin is much debated, the term Grand Staircase was first applied in 1924 by Charles Keyes in a paper written for the *Pan American Geologist*. The series

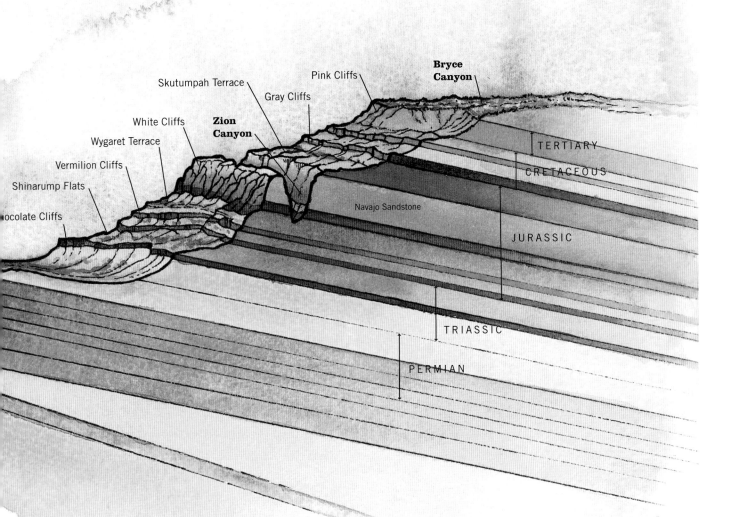

Skutumpah Terrace
Gray Cliffs
Pink Cliffs
Bryce Canyon
White Cliffs
Zion Canyon
Wygaret Terrace
Vermilion Cliffs
Shinarump Flats
Chocolate Cliffs

Navajo Sandstone

TERTIARY
CRETACEOUS
JURASSIC
TRIASSIC
PERMIAN

The bottom of the Grand Staircase commences at the top of the **Kaibab Uplift**, which correlates with and is in the same stratigraphic position as the highest bench of the Grand Canyon in Arizona.

The first riser above the bench is the poorly developed **Chocolate Cliffs** which consist of the upper red member of the Lower Triassic Moenkopi Formation capped by the Upper Triassic Shinarump Conglomerate Member of the Chinle Formation.

The next step is known as the **Shinarump Flats**. This bench is mostly developed on top of the resistant Shinarump Conglomerate Member and in the overlying, less resistant, Petrified Forest Member of the Chinle Formation.

The **Vermilion Cliffs** form the next riser. Zion National Park is found in this and the next two steps. The Vermilion Cliffs are composed of resistant red sandstone beds of the Lower Jurassic Moenave and Kayenta Formations.

The **Wygaret Terrace** forms the next step and includes the soft upper part of the Kayenta and the lower part of the Lower Jurassic Navajo Sandstone.

The imposing **White Cliffs** form the next riser and consist of the upper part of the Navajo Sandstone and part of the Middle Jurassic Carmel Formation. The bench on this riser is the **Skutumpah Terrace** built on the remaining soft parts of the Carmel Formation and the overlying Entrada Sandstone.

The **Gray Cliffs** are the next riser, and consist of a series of low cliffs formed by hard Cretaceous sandstone and less resistant shale beds.

The final riser, mostly north and west of Zion National Park, in Cedar Breaks National Monument and Bryce Canyon National Park, is formed by the **Pink Cliffs**. The Pink Cliffs consist of lower Tertiary limestones of the Claron Formation. The cliffs culminate as the Paunsaugunt Plateau, which is the uppermost bench or step of the Grand Staircase.

of topographic benches and cliffs that form the Grand Staircase step progressively up in elevation from southwest to northeast. The risers correspond to cliffs and the steps correspond to the broad benches, terraces, or plateaus in the staircase.

Plate Tectonics

Our understanding of the origin and history of the North American continent changed dramatically in the past several decades. This occurred because of scientific discoveries regarding our planet's structure and the forces that shaped it.

Evidence for an ancient Earth has been debated since the seventeenth century. But it wasn't until the beginning of the twentieth century that scientists offered an age of 4.6 billion years for the Earth. It is evident that during that vast period the Earth underwent significant change. Evidence includes exposed cores of ancient mountain ranges, ancient shorelines now perched on mountain tops, and an abundant marine fossil record now isolated far from modern oceans.

The key to understanding these enigmas is a theory that came into popular acceptance in the late 1960s. It was born of a desire to understand the origin of mountains, ocean basins, volcanoes, and earthquakes. Is there a controlling factor for these natural phenomena? Can it be observed in operation today? What part of the geologic record could be used to explain its existence?

ABOVE: In 1915, meteorologist Alfred Wegener first proposed the theory that continents moved great distances throughout Earth's history.

BELOW: Wegener's theory of continental drift was based partly on observations of the apparent puzzle-like fit of continents.

[24]

In 1915, a meteorologist named Alfred Wegener published a controversial, yet influential book entitled, *The Origin of Continents and Oceans.* Wegener proposed that continents had migrated great distances during Earth history. Ocean basins had opened and closed, creating tremendous changes on the exposed continents. His evidence for these events was a compilation of observations by scientists from Sir Issac Newton to American geologist Frank B. Taylor. This evidence was largely circumstantial, and included observations about the obvious puzzle-like fit of some continents, similarities in rock type and structures on continents widely separated by oceans, and an otherwise unexplainable distribution of certain fossil plants and animals.

Though his ideas made quite a splash in the scientific community, his theories were not widely accepted. This was largely due to his, and the rest of the scientific community's inability to explain the forces that drove continents, and the mechanism by which apparently solid earth moved. As a result, Wegener's theory of continental drift lay dormant for nearly one-half century. Other discoveries in the next several decades would ultimately support

EQUATOR

Zion National Park

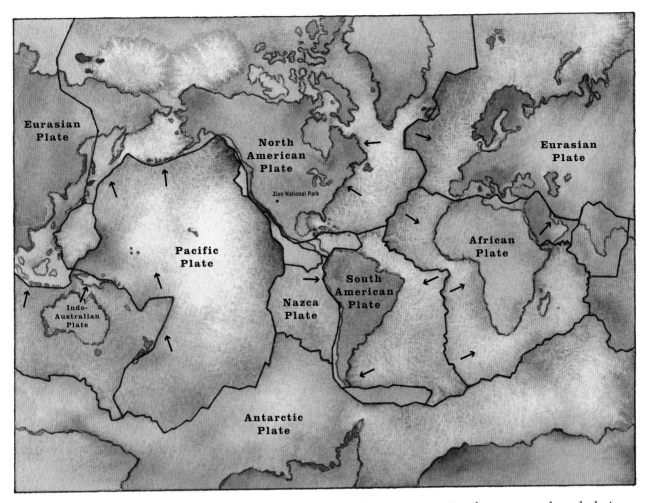

Eurasian Plate

North American Plate

Zion National Park

Eurasian Plate

Pacific Plate

African Plate

Indo-Australian Plate

South American Plate

Nazca Plate

Antarctic Plate

[25]

ABOVE: With the aid of SONAR, developed during World War II, scientists discovered that continents are passive passengers riding along on huge pieces of Earth's crust called oceanic plates. Arrows indicate direction of plate movement.

Wegener's hypothesis, but until more supporting data were gathered, their significance was not recognized.

In the post-Second World War era, wartime technologies were made available to science. One of the most significant of these was SONAR, which gave scientists the ability to remotely sense—or see—into areas that were previously unexplored. Most notable of these were ocean basins. With the aid of SONAR the ocean basins were mapped, revealing a submarine mountain system tens of thousands of miles long, making it the most extensive mountain range on Earth. This discovery led scientists to understand how continents move. The continents are passive passengers riding along on huge pieces of the Earth's crust called oceanic plates. These plates separate at the crest of the submarine mountain system, which literally results in ocean basins enlarging.

At the same time the submarine mountain ranges were discovered, ocean floor mapping revealed deep submarine canyons called oceanic trenches. The trenches were proposed as sites where oceanic plates dive back into our planet's interior. These discoveries, combined with decades of research in the world's ocean basins, help us understand the forces that shape the planet's surface.

DIVERGENT PLATE BOUNDARY

CONVERGENT PLATE BOUNDARY

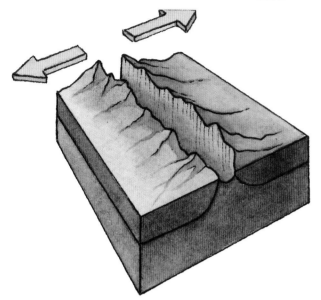

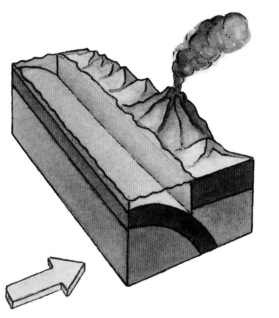

ABOVE: At divergent plate boundaries, new oceanic crust is formed as magma from the Earth's interior invades and the plates on either side are forced apart.

ABOVE RIGHT: At convergent plate boundaries, tectonic plates move toward each other and collide.

[26]

BELOW: At transform plate boundaries, two plates slide past one another laterally.

TRANSFORM PLATE BOUNDARY

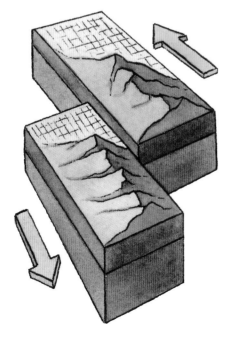

Earth's surface is divided into a series of rigid pieces called plates. These lithospheric plates move about the surface on a semi-molten, interior zone called the asthenosphere. The plates are bounded by ocean ridges, oceanic trenches, faults, and mountain belts. Plate boundaries are areas where volcanic activity, earthquakes, and mountain formation occurred repeatedly in the recent past. The evidence for this theory, now called plate tectonics, is overwhelming, universally accepted, and extremely important because it explains a variety of otherwise unrelated geologic features and events.

Geologists recognize three major types of plate boundaries (locations where lithospheric plates interact): divergent, convergent, and transform. Along these boundaries new plates are formed, consumed, and slide past each other. By studying present plate boundaries we have come to understand our planet's complicated history.

At **divergent** boundaries, new oceanic crust is formed from Earth's molten interior. These boundaries most commonly occur along ocean ridges, but occasionally occur beneath continents. As magma from the interior invades these regions, the plates on either side are forced apart. Divergence rates are variable, but the global average is approximately two to four inches (five to ten cm) per year, about the rate that fingernails grow. The Atlantic Ocean, for example, is split at its center by an oceanic ridge called the Mid-Atlantic Ridge. Consistent but slow movement on this ridge has only added about 100–160 feet (30–50 m) to the Atlantic since Columbus sailed in 1492. Although this rate of movement may seem insignificant, over the vast expanse of geologic time it has had major consequences.

At **convergent** plate boundaries, tectonic plates move toward each other and collide. As they do, mountains and volcanoes form. One side of the colliding plate is forced into the Earth's interior where it is recycled—a process called subduction. If the collision involves only ocean crust, it is an oceanic-plate-to-oceanic-plate convergence, and results in island-arc chains like the Japanese islands and the Aleutian chain.

When one colliding plate carries a continent into the collision, the oceanic plate material, which is always more dense than continental material, is forced into the Earth's interior. The result is volcanic and earthquake activity that produces mountain ranges on the continent side similar to the modern Cascade and Andes Mountains. Further inland the collision may cause large blocks of a continent to rise in relation to their surroundings. The uplift of the Colorado Plateau, which led to the sculpting of Zion National Park, is related to a collision between the North American and the Pacific Plates.

Occasionally, two plates slide past one another laterally, creating a **transform** plate boundary. This type of plate boundary rarely produces volcanic activity, but is responsible for some very significant earthquake activity. The San Andreas Fault in California is probably the best-known example of a transform plate boundary. The northward movement of the Pacific Plate grinding against the North American continent continues to produce periodic earthquake activity.

When our planet's geologic history is viewed with an understanding of plate tectonics it's easy to see that some of the Earth's most significant features—the Himalayas, the Cascades, and the San Andreas Fault—are clearly related to this global phenomenon. Even old mountain ranges like the Appalachians are the result of continental collision.

The migration of continents through geologic time explains things like global climatic variations reflected in the fossil record of the Zion region. Worldwide fluctuations in sea level are directly related to rates of spreading at divergent plate boundaries. These factors combine to illuminate our understanding of the complicated geologic history of Zion National Park.

Reflections after a storm on the park's east side.

"The history of man has been recorded in the tomes of libraries only since the invention of letters, but a much longer period anterior to that is recorded on the leaves of the book of rocks."

—JOHN WESLEY POWELL, *Forum*, Vol. 8, January 1890

The Geologic Record

O UR ABILITY TO RECONSTRUCT the history of this planet is tied to our understanding and interpretation of the rock record. Evidence is preserved in the origins, distributions, and interrelationships of rocks. We will explore Zion's rock record in terms of the materials that make it up, the time frame in which these materials accumulated, and the relationship of these earth materials to each other.

OPPOSITE PAGE: Sunset above Phantom Valley casts an other-worldly light across the rock and autumn foliage atop Zion's West Rim near Horse Pasture Plateau.

ABOVE: A mesa capped by the Temple Cap Formation along the Zion-Mt. Carmel Highway on the park's east side.

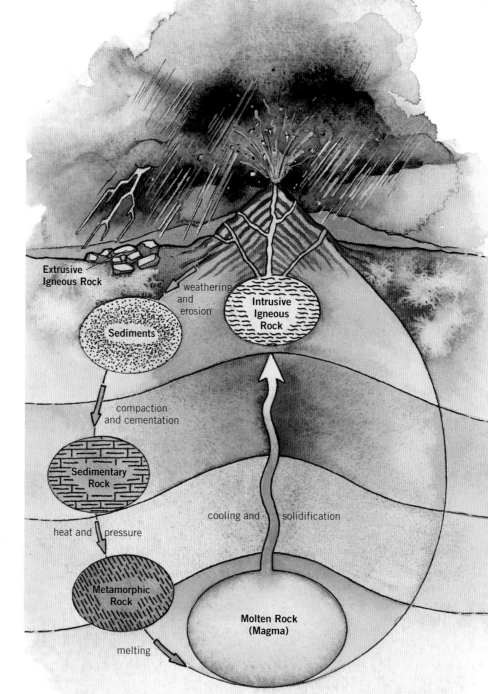

[30]

Earth Materials

One way of viewing the interaction between earth materials and the Earth's
internal and external processes is to consider their relationships to one another. A graphical representation of this is called the rock cycle. This concep-
tual framework allows us to relate the three major rock types to each other,
to surface processes such as weathering and transportation, and to internal
processes such as magma generation and mountain building.

Magma, molten rock material below the Earth's surface, is the building
block of earth materials. When this magma breaks through to the surface
it is called lava. Geologists recognize two major categories of **igneous rock:**

volcanic (solidifying at the surface) and plutonic (cooling into a solid within the Earth). Although there are no plutonic rocks exposed in Zion, and only relatively small, recent exposures of volcanic rock, we will see that igneous rocks contribute to the geologic history of Zion National Park.

Any rocks exposed at the Earth's surface are subjected to mechanical and chemical processes that break them down, yielding the raw material for **sedimentary rock**. The combination of all processes that cause rocks to break apart is called weathering. Weathered material is carried by gravity and running water to a low point where it accumulates. The accumulation of loose, weathered earth materials in low places like valleys and ocean basins by gravity, ice, wind or running water, is called deposition. The accumulating loose sediment is compacted by each addition of more sediment above. With enough time, and circulating underground water, the buried, accumulated sediment is transformed from loose material to solid sedimentary rock. The kind of rock—sandstone, shale, or limestone—is determined by the raw materials, the climate in which it formed, and the depositional environment. Most of the story of Zion National Park is preserved in sedimentary rocks deposited over the past several hundred million years.

Metamorphic rock results from the transformation of other rock. These transformations, which are the result of heat, pressure, and chemical change, include everything that happens to the rock up to the point of melting. There is no metamorphic rock visible in Zion National Park, but this important group of rocks forms the foundation for the towers of Zion.

[31]

All rock types are important in understanding the geologic record: igneous rocks record activity related to diverging and converging plates. Metamorphic rocks reveal evidence of the dynamic processes within our planet.

Although sedimentary rocks only account for 5 percent of the Earth's outer ten miles (15 km) they make up 75 percent of the rocks exposed at the planet's surface. They cover most of the seafloor and about two-thirds of the continents.

Sedimentary rocks not only record climates and environmental conditions, they may contain fossils that tell us how particular environments, and our planet, evolved through history. The appearance of fossils also provides a means of determining the age of rock. Finally, in the Colorado Plateau, the layers of sedimentary rock are extensive and traceable. The correlation of sedimentary rocks of similar age and origin exposed in widely separated locations is one way we unravel our planet's history.

Geologic Time

In geology, time is examined in two different ways: the chronological order of events (relative time), and actual ages (absolute time). Relative time compares two or more rocks or events to determine which is older and which is younger. Just as you can estimate the relative ages of people because you know the basic body characteristics of people of various ages, geologists understand certain characteristics that help determine the relative ages of rocks and geological features. Those characteristics include superposition (the oldest layers are on the bottom), fossil succession (life on Earth has changed in a definite order through time), cross-cutting relationships (faults must be younger than the rocks they cut), and the principle of inclusions (fragments of sand in a sandstone must be older than the sandstone that contains them).

Absolute time is measured in years, and establishes approximately how many years ago a specific feature developed or an event occurred. To determine the absolute age of an individual, you would need a birth certificate or driver's license. The Earth contains no such easily readable record. Geologists use a variety of sources to determine the age of geologic events and materials including: radiometric dating (for very old events), tree-ring dating (for more recent events), and the study of lichen colony growth (for events in the past 9,000 years). These and other dating methods are used to interpret the history of Zion National Park and the Colorado Plateau.

Relative and absolute geologic time have been combined to establish a calendar of Earth history. Just as our modern Roman calendar is divided into periods of varying lengths (days, weeks, months, years, decades, centuries, etc.), geologic time can be subdivided into components of different lengths.

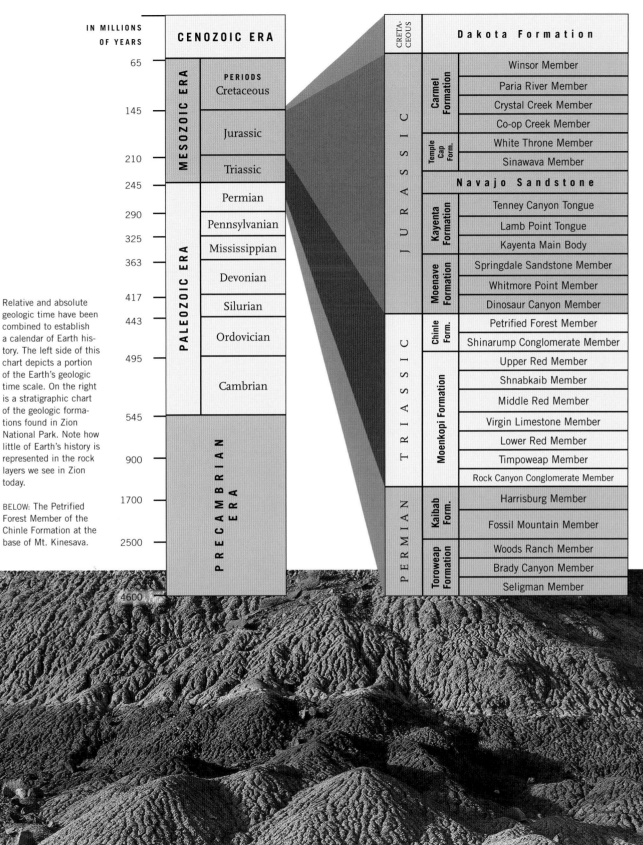

IN MILLIONS OF YEARS

CENOZOIC ERA

Era	Periods	Million years
		65
MESOZOIC ERA	Cretaceous	
		145
	Jurassic	
		210
	Triassic	
		245
PALEOZOIC ERA	Permian	
		290
	Pennsylvanian	
		325
	Mississippian	
		363
	Devonian	
		417
	Silurian	
		443
	Ordovician	
		495
	Cambrian	
		545
PRECAMBRIAN ERA		900
		1700
		2500
		4600

Relative and absolute geologic time have been combined to establish a calendar of Earth history. The left side of this chart depicts a portion of the Earth's geologic time scale. On the right is a stratigraphic chart of the geologic formations found in Zion National Park. Note how little of Earth's history is represented in the rock layers we see in Zion today.

BELOW: The Petrified Forest Member of the Chinle Formation at the base of Mt. Kinesava.

CRETACEOUS	**Dakota Formation**	
JURASSIC	Carmel Formation	Winsor Member
		Paria River Member
		Crystal Creek Member
		Co-op Creek Member
	Temple Cap Form.	White Throne Member
		Sinawava Member
	Navajo Sandstone	
	Kayenta Formation	Tenney Canyon Tongue
		Lamb Point Tongue
		Kayenta Main Body
	Moenave Formation	Springdale Sandstone Member
		Whitmore Point Member
		Dinosaur Canyon Member
TRIASSIC	Chinle Form.	Petrified Forest Member
		Shinarump Conglomerate Member
	Moenkopi Formation	Upper Red Member
		Shnabkaib Member
		Middle Red Member
		Virgin Limestone Member
		Lower Red Member
		Timpoweap Member
		Rock Canyon Conglomerate Member
PERMIAN	Kaibab Form.	Harrisburg Member
		Fossil Mountain Member
	Toroweap Formation	Woods Ranch Member
		Brady Canyon Member
		Seligman Member

[33]

ABOVE: Zion's West Temple is a beautifully symmetrical tower of Kayenta and Navajo Sandstone Formations topped by the Temple Cap Formation.

BELOW: An example of sharp bedding planes in Kolob Canyons.

Stratigraphic Concepts in Zion National Park

Since so much history of Zion National Park and the Colorado Plateau is contained in layered rocks, it is important to understand a few basic principles relating to them. The study of layered rocks and their relationships is called stratigraphy. Stratigraphy requires an understanding of both vertical and lateral rock relationships.

For example, the most fundamental aspect of the vertical relationship is where one rock layer stops and the next one begins. This boundary between layers is called a bedding plane. Bedding planes can be gradational or sharp. A gradational bedding plane slowly transitions upward into another rock type, and represents a gradual change in the conditions under which the sediment accumulated. A sharp bedding plane marks boundaries in layered rocks where the rocks above and below are very different. Abrupt changes in the composition of layered rocks suggest either rapid changes in sediment input, or a period of exposure and erosion at the surface, followed by later depositions.

Lateral relationships are important because they trace the boundaries of layered rocks and determine how extensive a given deposit is. Lateral relationships also allow us to compare rocks from different locations and decide if two different rock exposures are the same age (a process called correlation).

RIGHT: These sharp bedding planes mark the contact between beds in the Moenave Formation.

The Rock Record of Zion National Park and the Colorado Plateau

For the purpose of discussion, Earth history can be divided into three time periods: the geologic history that precedes the rocks exposed in Zion National Park 245 million years ago; the geologic history of Zion National Park as told by rocks of the Mesozoic Era (245 to 66 million years ago); and finally, the geologic history contained in Cenozoic Era rocks (younger than 66 million years).

The vast majority of geologic time (over four billion years of history), is not visible in Zion National Park. We can, however, catch a glimpse of it if we examine the oldest rocks revealed at river level in the Grand Canyon's Inner Gorge. These rocks, called the Granite Gorge Metamorphic Suite, are part of the Colorado Plateau's basement, and are approximately two billion years old. While not the oldest rocks on Earth, they are the oldest visible in the region. Buried beneath thousands of feet of younger sedimentary rocks, these metamorphic rocks record a period when North America was growing due to plate collisions. When these basement rocks were formed the face of our globe was very different than it is today. The continents were smaller and the configuration of the oceans was different. The area in which we now find these basement rocks was then sea floor, lying between a younger North America and a series of off-shore volcanic islands. These volcanic islands eventually collided with North America. Tremendous heat and pressure—and later invasion by magma—disguised the original character of the volcanic island rocks, but they give the trained eye a glimpse into the region's complex history.

Lying above a sharp erosional bedding plane on the metamorphic basement are horizontally layered rocks of the Paleozoic Era (540 to 245 million years ago). Some of the best rock exposures formed during this time period are found, again, in the walls of the Grand Canyon and record a global event at the beginning of the Paleozoic. It was a period of **transgression**: sea level began to rise and moved onto the land, covering many parts of continents with shallow waters. It is hard for us to envision how this could occur today, since most land mass is high and dry, and the oceans are trapped in deep basins, but a few shallow, intercontinental seas remain such as Canada's Hudson Bay. These intercontinental seas were the sediment source for most of the 300 million years of Paleozoic time. The sequence of rock types actually records the shore's gradual migration inland, and shallow ocean water covering former shorelines. The sequence of transgression is basal sandstone (the shoreline) overlain by shale (the accumulated mud of the offshore marine environment), which, in turn, is overlain by limestone (limy sediments of quieter, deeper marine waters). The environmental interpretation for each environment is supported by fossil organisms preserved in each rock type.

Much of the rest of the 300-million-year Paleozoic rock record is the result of rising and falling sea levels. We recognize two kinds of sea level change: worldwide change, and more localized change due to the rise and fall of local landmass. What are the possible causes of sea level change? One is increasing or decreasing polar ice caps. Since the total volume of water on Earth is relatively fixed, ocean levels are directly tied to ice cap size. When the permanent ice bodies at the poles are large, the ocean levels must drop. During global warming periods, the amount of water in the permanent ice caps decreases, causing sea level to rise and coastal areas to flood. Paleozoic sea level changes, however, are more likely related to an increase or decrease

[37]

LEFT: A gnarled juniper frames the view of Zion's East and West Temples. Here the high cliffs of spectacularly cross-bedded Navajo Sandstone represent the largest preserved coastal and inland dune system in the geologic record of North America.

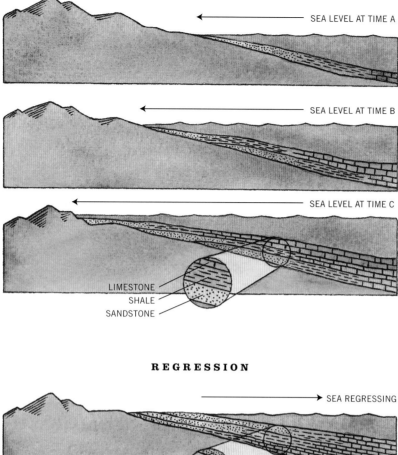

TRANSGRESSION

SEA LEVEL AT TIME A

SEA LEVEL AT TIME B

SEA LEVEL AT TIME C

LIMESTONE
SHALE
SANDSTONE

REGRESSION

SEA REGRESSING

SANDSTONE
SHALE
LIMESTONE

Transgression occurs when sea level begins to rise and the sea moves onto the land, covering it with shallow water. A sequence of rock types records the gradual migration of the shore inland with basal sandstone overlain by shale overlain by limestone.

During periods of rising sea levels the Paleozoic Colorado Plateau received sediment from the seas. When sea level fell, known as regression, the region was exposed to erosion. During regression the sequence of rock types reverses.

in the volume of the ocean basins themselves, related to variations in the rate of sea-floor spreading. For example, rapid sea-floor spreading increases the volume of the mid-ocean ridge, and decreases the volume of the ocean basins, causing sea level to rise.

During periods of sea level rise, the Paleozoic Colorado Plateau received sediment from the seas. When sea level fell (known as **regression**), the region was exposed to erosion. Although the volume of Paleozoic rock on the Colorado Plateau is huge, this pattern of sea level fluctuation causes significant gaps in the rock record. Almost as if someone had torn large chapters out of a history book, the missing rock layers are literal gaps in our understanding of Paleozoic Earth history. These missing pages, called unconformities, are a significant part of the Colorado Plateau's history.

LATE PALEOZOIC GEOLOGIC HISTORY OF ZION NATIONAL PARK

Toward the end of Paleozoic history, the North American continent collided with a large landmass on its eastern margin. This collision was the final event in the Paleozoic and resulted in the formation of the Appalachian Mountains on the eastern edge of North America. By the end of the Paleozoic, all of the major continents had collided to form one large land mass known as Pangaea. The formation of Pangaea was followed by a global oceanic regression, as recorded in a sharp bedding plane between the Paleozoic and Mesozoic rocks of the Colorado Plateau. The end of the Paleozoic also represents the largest extinction of marine life in Earth history. It is estimated that 90 percent of all marine invertebrate species ceased to exist. This catastrophic extinction in the marine realm was probably the result of a reduction in living areas for these animals due to global regression, and a change in the salinity of the oceans due to global desert conditions on land.

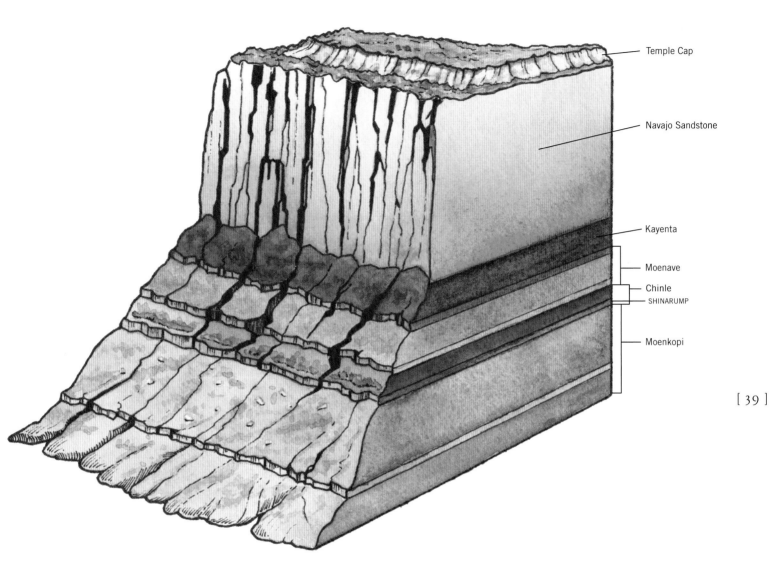

Temple Cap

Navajo Sandstone

Kayenta

Moenave

Chinle

SHINARUMP

Moenkopi

[39]

A cross section of the Zion region reveals its geologic history through the Mesozoic Era, from the Triassic Period through the Jurassic and into the Cretaceous Periods.

The oldest sedimentary bedrock seen in Zion National Park is located in Kolob Canyons. Here the Early Permian Toroweap Formation, a tan to gray shallow-marine limestone, can be seen at the base of the Hurricane Cliffs. Immediately above the Toroweap Formation are the tan to gray, shallow-marine limestones of the Kaibab Formation. Deposition of these rocks occurred on the western edge of Pangaea when the North American plate lay in an equatorial position. During the Late Paleozoic, this portion of Pangaea experienced alternating marine transgression and regression, exposing broad, coastal sabkhas (very flat surfaces near sea level that experience high evaporation rates, for example, the modern Arabian Peninsula.) The evidence for the marine origin of these rocks includes a variety of fossilized invertebrate organisms.

[40]
This panoramic northward view of the mouth of Zion Canyon highlights Mt. Kinesava on the left and the east rim of Zion on the right. The Early Jurassic Navajo Sandstone reaches its greatest thickness in Zion National Park, exceeding 2,000 feet (610 m).

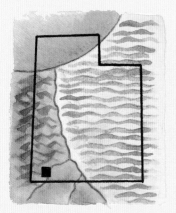

**Early Jurassic
Moenave Formation**

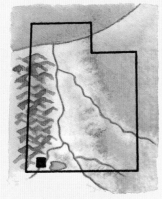

**Early Jurassic
Kayenta Formation**

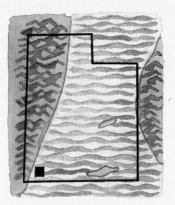

**Early Jurassic
Navajo Sandstone**

**Middle Jurassic
Temple Cap and Carmel
Formations**

MESOZOIC GEOLOGIC HISTORY OF ZION NATIONAL PARK

OPPOSITE: These simplified paleo-geographic maps show the depositional environments of the Utah and Zion area over time. Modified from DeCourten (1998).

In the first diagram the Wingate Sandstone was the first of two large dune fields to develop in Early Jurassic time, while to the southwest the Moenave Formation was deposited in fine grained clays and sandstones accumulated in a variety of river, lake, and flood-plain environments.

The second diagram shows the Kayenta Formation appearing as a result of stream deposited sands and muddy flood-plain deposits. In the third diagram the Zion region was engulfed in desert conditions resulting in the gradual southwestern expansion of the Navajo Desert, an immense sand sea that covered at least 150,000 square miles (241,400 km²) of western North America.

The fourth diagram shows the returning seaway which came as far south as the Zion area and deposited the Carmel Formation.

A ten-million year unconformity (gap in the rock record due to erosion) separates Paleozoic rocks and Mesozoic rocks of the Triassic Period. It represents a period of exposure and erosion of the latest Paleozoic rocks. The first period of the Mesozoic, the Early Triassic, represents a continuation of Permian geography, with North America occupying an equatorial position. The Moenkopi Formation marks the renewal of sediment accumulation on the western margin of Pangaea. The rocks of the Moenkopi Formation record three shallow-marine transgressions (the Timpoweap, Virgin Limestone, and Shnabkaib Members), separated by three regressions (the informal lower, middle, and upper red members, respectively). Locally, the base of the Moenkopi Formation is the Rock Canyon Conglomerate Member, a pebbly conglomerate exposed in the Hurricane Cliffs outside the park. The pebbles in this unit came from the underlying Kaibab Limestone, while an uplift to the southeast supplied finer-grained sediments. The minerals in the Moenkopi attest to the arid climate of what would become southern Utah during Early Triassic time. Poorly circulated, mineral-rich waters evaporated, depositing beds of gray gypsum and siltstone, common components of the Moenkopi. The Moenkopi Formation is exposed at the top of the Hurricane Cliffs along the Hurricane Fault-zone (see diagram pg. 49) of the Kolob Canyons, and in the lower portions of Coalpits and Huber Washes.

ABOVE: The Petrified Forest
Member of variegated gray, purple
and white shale weathered into
bare clay hills is the upper member of the Chinle Formation, seen
here at the base of Mt. Kinesava.

[42]

RIGHT: A medley of sandstone
forms illustrates the true artistry of
nature working in the medium of
sedimentary rock.

Moenkopi deposition ceased by the end of the Early Triassic and an interval of exposure and erosion lasting twenty million years followed. This unconformity is overlain by the Shinarump Conglomerate, the basal member of the Chinle Formation. The Chinle Formation marks the beginning of Late Triassic deposition, and a change from mostly shallow-marine to continental sediment deposition on the Colorado Plateau. Consisting of two members in the Zion National Park area—the Shinarump Conglomerate and Petrified Forest Members— the Chinle Formation was first named and described near Chinle, Arizona. The sandstone, pebbly sandstone, and pebbly conglomerate of the Shinarump Conglomerate was deposited in stream channels and can be seen as a prominent rock layer capping areas such as the Rockville Bench above the town of Rockville in the park's southwest corner. The Petrified Forest Member was deposited in flood plains, lakes, and stream channels. This rock unit is commonly covered by landslide deposits, but where revealed, its colors are bright, banded, and distinctive. The Chinle is perhaps most famous for its petrified wood found in the Petrified Forest Member, but it also contains amphibian and reptile tracks, and freshwater fossil forms.

Following Chinle deposition, the region became more arid as evidenced by dune and playa (dry) lake deposits at the top of the formation. This pattern of desert-like conditions continued into the Jurassic Period. A five- to ten-million-year period of non-deposition and erosion resulted in a Late Triassic-Early Jurassic unconformity. The arid conditions that began in the latest Triassic continued throughout the Jurassic. These climate conditions developed because of changes in North American geography due to the breakup of Pangaea.

Sediment deposits of the Moenave Formation cap the Chinle Formation. These fine-grained sandstones accumulated in a variety of river, lake, and flood-plain environments. The Moenave Formation is divided into the gentle slopes of the Dinosaur Canyon and Whitmore Point Members and the overlying cliff-forming Springdale Sandstone Member. The slope-forming lower units are primarily composed of fine-grained sandstones and silty sandstones. The Springdale Sandstone Member forms the first significant cliff below the Navajo Sandstone, and was deposited in river channels. It is best exposed in the cliffs immediately above the town of Springdale, and can be readily identified there.

The Moenave Formation eventually gives way to the stream-deposited sands and muddy flood-plain deposits of the Kayenta Formation. The Kayenta

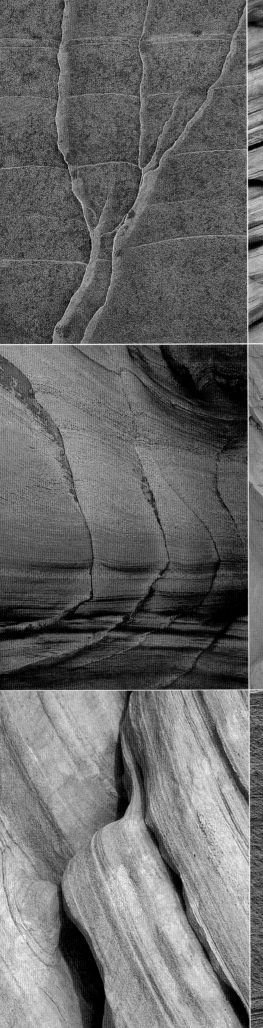

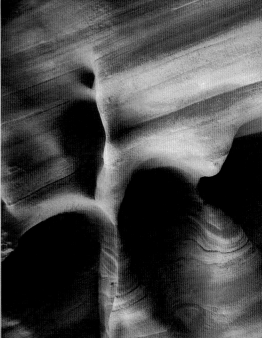
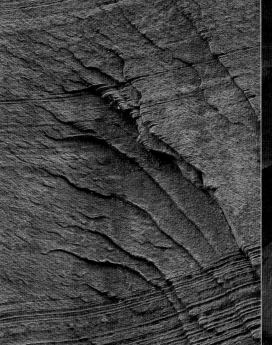
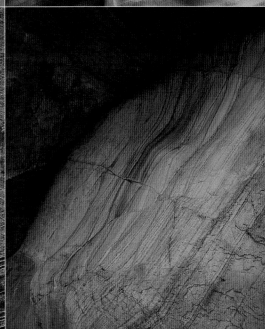

ABOVE: The Temple Cap Formation is seen here as the literal cap on Zion's West Temple.

BELOW: Zion's towers and domes were once part of an immense sand dune desert on the scale of the modern Sahara.

Formation consists primarily of siltstones and fine-grained sandstones and forms deep red, steep slopes between the cliffs of the Springdale Sandstone below and Navajo Sandstone above. It contains fossils, particularly burrows and root impressions that support its suspected stream/flood-plain origin.

As plate motions of the Jurassic continued to move North America northward, mountains in Nevada and California created a rain shadow over the Colorado Plateau. By Middle Jurassic time, the Zion National Park region was engulfed in desert conditions resulting in a gradual south-southwestern expansion of the Navajo Desert (a huge dune field on the scale of the modern Sahara) until it overtook the Kayenta flood-plain. The transition is amazingly recorded in the gradational contact between the Kayenta Formation and the overlying Navajo Sandstone.

The Early Jurassic Navajo Sandstone reaches its greatest thickness in Zion National Park, exceeding 2,000 feet (610 m). The Navajo desert environment spread sand over an area of about 150,000 square miles (390,000 km²). The high cliffs of spectacularly cross-bedded Navajo Sandstone represent the largest preserved coastal and inland dune system in the geologic record of North America. The Navajo Sandstone is well known for its internal uniformity, consisting of well-rounded, fine to medium sand grains. The sand was likely recycled from Paleozoic and Triassic sandstones to the north, possibly as far north as Alberta, Canada, and from local sources. In Zion National Park measurements of single cross-bed sets suggest a minimum dune relief of over 60 feet (18.3 m).

Late in the Middle Jurassic, sea level began to rise once again and a sea transgressed southward from Idaho and Wyoming. During this transgression the upper Navajo Sandstone underwent a period of erosion. In southwestern Utah, this brief break from desert conditions is recorded in the Sinawava Member of the Temple Cap Formation. Lying in a shallow depression atop the

Navajo Sandstone, the Sinawava Member is composed of bright red, thinly bedded mud. This brief period of marine transgression was followed by a brief return to desert conditions, and the White Throne Member of the Temple Cap Formation was deposited on these exposed marine beds. The position of this unit on the Navajo forms a cap on the many temples of Zion National Park, hence the name Temple Cap Sandstone. Like the Navajo Sandstone, the Temple Cap Formation was eroded flat by a subsequent transgression.

Plate collision and subduction off the west cost of North America created eastward-directed compression of the continent. In the area of Zion National Park a basin was created by this compression, and with the rising of sea level, eventually marine conditions returned, as evidenced by the Carmel Formation overlying the Temple Cap Formation. The Carmel Formation is composed of many different rock types due to rapid changes in basin conditions. The Carmel Formation in Zion National Park consists of four members, in ascending order: Co-op Creek Limestone, Crystal Creek, Paria River, and Winsor Members. The Carmel Formation appears as gray, fossil-rich hills atop the Temple Cap Formation and is best observed near the park's east-, northeast-boundary.

By late Middle Jurassic time, most of Utah was high and dry. In southwestern Utah, this broad, gentle uplift was enough to produce a long period of modest erosion. The result is that there are no rocks of late Middle Jurassic to middle Early Cretaceous age preserved in southwestern Utah. By late Early Cretaceous time, pebbly conglomerates, sandstones, and mudstones were deposited on the earlier eroded surface. Within the park, Cretaceous rocks are limited to exposures at the top of Horse Ranch Mountain in the Kolob Canyons area. However, a more diverse suite of Cretaceous rocks, several thousand feet thick, can be found north and east of Zion National Park.

Within Zion National Park, Cretaceous rocks are limited to exposures at the top of Horse Ranch Mountain in the Kolob Canyons area. This is the park's highest point at 8,726 feet (2,660 m) above sea level.

[45]

Cenozoic Geologic History of Zion National Park

Tertiary-age sedimentary and volcanic rocks once covered Zion National Park, but have been removed by erosion. The river- and lake-deposited sedimentary rocks of the Claron Formation form the base of the Tertiary unit throughout southwest Utah. Though these rocks are beautifully exposed at Cedar Breaks National Monument and Bryce Canyon National Park, they are not present in Zion National Park. The extensive middle Tertiary explosive volcanic deposits so common north of the park have also been stripped away by erosion.

As we examine Quaternary evidence, moving ever closer to the present in our examination of the geologic history of Zion National Park, we discover that the past two million years have wrought significant geologic changes. One significant group of geologic events is evidenced by the presence of basaltic lava flows and cinder cones located within, and adjacent to the park.

Zion National Park lies within the Western Grand Canyon basaltic field. Most volcanic flows in this field are less than seven million years old, and the youngest is less than 500 years old. The basalt flows of Zion National Park are fairly young and range in age from 1.4 million to an estimated 100,000 years old. There are parts of six lava flows within park boundaries. The flows are generally less than 40 feet (12 meters) thick, but may be several hundred feet thick where they filled canyons. These basaltic lava flows not only provide a contrast to the pervasive red-rock scenery, they allow important insights into the uplift and erosional history over the past two million years. Each basalt flow was emplaced in a geological instant, can flow for many miles, and is resistant to erosion. These flows drape existing features in a protective blanket that resists the continuing erosion in the region. The basalt, which originally flowed down the valley floors and canyons, protects what lies beneath it while the adjacent sedimentary rocks are stripped away by erosion. The result is an inversion of topography (see pg. 70). Former valley floors become basalt-capped mesas, creating not only spectacular scenery, but also a means of calculating rates of downcutting erosion in the Zion area. These basalt flows have also dammed rivers and streams, resulting in temporary lakes and ponds and their associated sediments. In addition, they have caused streams to change resulting in the formation of new erosional features.

Landslides occur periodically in the Zion area. These events can also dam rivers and streams, creating lakes and ponds behind them. Part of the park's recent geologic history includes evidence of significant landslides. A recent

TOP: Crater Hill, west of Coal Pits Wash on the park's western boundary, is a relatively young cinder cone marking one of the last episodes of volcanic action in Zion.

ABOVE: Basalt beds visible along the Kolob Terrace Road.

[46]

landslide event may serve to illustrate this process and its effects. On April 12, 1995, at about 9:00 P.M., a slide occurred in the lower portion of Sand Bench in Zion Canyon, just north of Canyon Junction. This landslide dammed the North Fork of the Virgin River, forming a pond about 20 feet (6 m) deep. The ponded water eventually overflowed the dam, but no flooding occurred downstream. However, in its attempt to escape impoundment the river eroded its east bank, removing a 600-foot (180-m) section of the Zion Canyon Scenic Drive. These events stranded 300 people upstream at Zion Lodge for approximately 36 hours. The 1995 landslide mass was roughly 500 feet (150 m) long and 150 feet (45 m) wide, and included 110,000 cubic yards (84,000 m³) of debris. The probable cause of the slide was higher than usual precipitation that year (189 percent of the water-year average for that period), which resulted in an over-saturated sediment mass on lower Sand Bench.

[47]

The 1995 Sentinel Slide dammed the North Fork of the Virgin River along the main canyon drive for a short time, forming a pond about twenty feet deep.

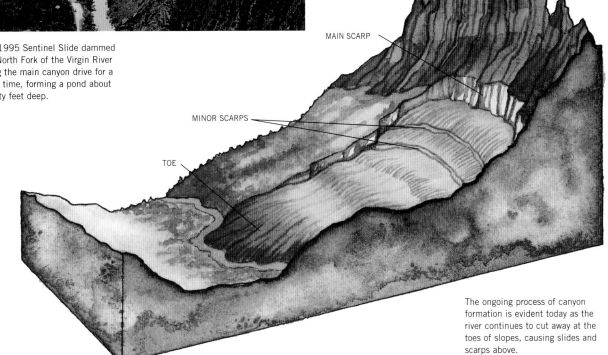

MAIN SCARP

MINOR SCARPS

TOE

The ongoing process of canyon formation is evident today as the river continues to cut away at the toes of slopes, causing slides and scarps above.

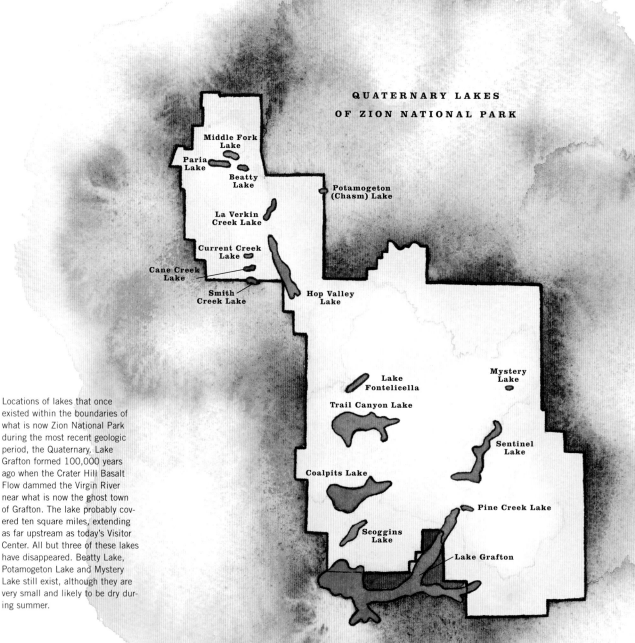

QUATERNARY LAKES
OF ZION NATIONAL PARK

Middle Fork
Lake

Paria
Lake

Beatty
Lake

Potamogeton
(Chasm) Lake

La Verkin
Creek Lake

Current Creek
Lake

Cane Creek
Lake

Smith
Creek Lake

Hop Valley
Lake

Mystery
Lake

Lake
Fontelicella

Trail Canyon Lake

Sentinel
Lake

Coalpits Lake

Pine Creek Lake

Scoggins
Lake

Lake Grafton

Locations of lakes that once existed within the boundaries of what is now Zion National Park during the most recent geologic period, the Quaternary. Lake Grafton formed 100,000 years ago when the Crater Hill Basalt Flow dammed the Virgin River near what is now the ghost town of Grafton. The lake probably covered ten square miles, extending as far upstream as today's Visitor Center. All but three of these lakes have disappeared. Beatty Lake, Potamogeton Lake and Mystery Lake still exist, although they are very small and likely to be dry during summer.

[48]

Because of the ability of lava flows and landslides to create temporary lakes and ponds, the canyons of Zion National Park have periodically held impounded water. The sediments that accumulate behind such dams contain fossils of various kinds that provide a snapshot of environmental conditions at the time the sediment accumulated. There are sediment deposits associated with at least 14 former lakes in the park. The largest of these, Lake Grafton, formed 100,000 years ago when the Crater Hill Basalt Flow dammed the Virgin River near what is now the ghost town of Grafton. Lake Grafton probably covered about ten square miles (26 km²) and may have extended upstream as far as the Visitor Center. Subsequent erosion removed most of the sediments associated with this and other park lakes.

The most famous of Zion's temporary lakes is Sentinel Lake, formed upstream of the Sand Bench landslide area. This lake existed at least 7,000 years ago and extended from the Court of the Patriarchs upstream nearly to the Temple of Sinawava. The lake was at least 200 feet (61 m) deep at one point and probably held water year-round. Although the Virgin River has since eroded most of its sediment, the thin, gray-clay and yellow-sand, deposits of Sentinel Lake can be found exposed in several side canyons near Zion Lodge.

The Structure of the Land

Zion National Park lies at the western edge of the Colorado Plateau, near the transition to the Basin and Range physiographic province. The Colorado Plateau is a tectonically stable region composed mostly of horizontal sedimentary strata with only minor, local disruptions. In contrast, the Basin and Range province is comprised of block-faulted north-trending uplifted mountain ranges with intervening down-dropped valleys. The transition zone between them is a mixture of sedimentary strata and structures found in both.

In the Zion area, the western Colorado Plateau and transition zone correspond to the leading edge of a Jurassic to Early Tertiary compressional event that created the Kanarra Anticline. This broad, upward-arching feature affects the otherwise horizontal sedimentary layers of the region and has subsequently been dissected by canyons. The Kanarra Anticline is located in the Kolob Canyons area.

The transition zone and western Colorado Plateau also include several major fault zones. Zion National Park lies in a structural block bounded on the west by the Hurricane Fault zone and on the east by the Sevier Fault zone. The Hurricane Fault zone is the only major fault system found within park boundaries (Kolob Canyons area); however, there are several minor faults exposed as well. The Hurricane Fault zone is an active, north-trending struc-

[49]

Zion National Park lies in a structural block bounded on the west by the Hurricane Fault zone and on the east by the Sevier Fault zone. The Hurricane Fault zone extends more than 155 miles (250 km) from south of the Grand Canyon north to Cedar City, Utah.

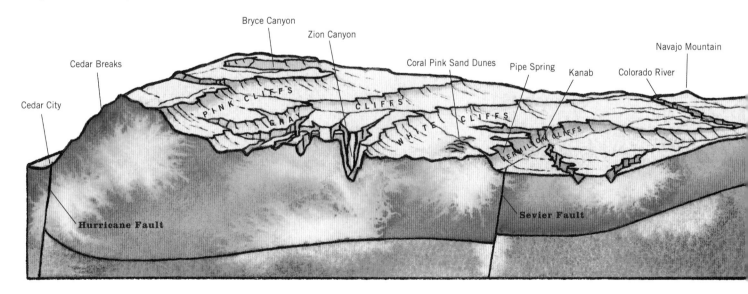

ture that displays significant evidence of movement in the last two million years. It extends from south of the Grand Canyon north to Cedar City, a distance of over 155 miles (250 km). The average slip rate of the fault, measured near the Kolob Canyons area, is about 16 inches (40 cm) per thousand years, for the last million years. The date of initial Hurricane Fault movement is controversial, but it probably started between 11 and 3 million years ago.

Although there are few faults in Zion National Park, fracture systems, called joints, are very well developed. These joint systems provide pathways for water runoff and control orientation of the present canyon system. The most prominent joints in the park are north-trending, nearly vertical, and widely spaced. In contrast, some joints form parallel to the rock faces and are responsible for arch and window formation. They also explain the rounded appearance of many Navajo Sandstone surfaces. Short, shallow joints, such as those exposed at Checkerboard Mesa, are the result of local expansion and contraction of the rock, and are probably due to changes in temperature and moisture.

OPPOSITE RIGHT: Aerial view of the Kolob finger canyons in the park's northwest section.

BELOW: Checkerboard Mesa, one of the park's icons, derives its name from its vast array of horizontal and vertical grooves. Sandstone bedding structures form the horizontal grooves while vertical grooves are formed by shallow fractures caused by expansion and contraction of the rock surface.

[50]

"Especially in the narrow canyons that stand like fossilized floods, this is where water truly gets about its business, revealing intimate details about the life of water. Walking in these places is like tracking an animal, studying the finesse of its prints."

—CRAIG CHILDS, *The Desert Cries*, 2002

The various tributaries of the Virgin River erode their channels by abrasion, hydraulic lifting, and dissolution. Here, the Left Fork of North Creek has carved a subway-like channel through the Navajo Sandstone.

Water

and the Geology of Zion National Park

WATER IS AN IMPORTANT COMPONENT in the past and current story of Zion National Park. Climate is the controlling factor in precipitation, the source of the park's water resources. Average precipitation equals 14.4 inches (36.6 cm) per year. The maximum reported annual precipitation in Zion National Park, recorded in 1978, was 25.9 inches (65.8 cm). Precipitation during the driest year on record, 1956, was reported at 3.20 inches (8.1 cm). This extreme variation is not, however, unusual for the semiarid climate of Zion National Park.

A hiker cools off in a seasonal waterfall in Kolob Creek.

ABOVE: A storm passes over Bridge Mountain, visible from the Zion Human History Museum. The monsoon season in Zion is typically July through August.

LEFT: Sunset casts a magical light across the gently flowing Virgin River.

OPPOSITE BOTH IMAGES: The effects of water's erosive nature are seen throughout Zion on sandstone surfaces. Summer rains collect in potholes creating mini-ecosystems in the slickrock.

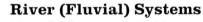

An intrepid hiker wades through the cool waters of the Zion Narrows.

Precipitation falling to Earth is called meteoric water, and it enters the park's hydrologic system in one of two ways: from summer thundershowers, locally known as monsoons, or winter frontal storms. Summer thundershowers generally flow or run off along existing drainage networks, eventually exiting the park via the Virgin River. Winter frontal storms are usually confined to the highest elevations and have a better opportunity to infiltrate porous sediments and rock, becoming part of the park's groundwater system. Let's examine these two important components of Zion National Park hydrology.

River (Fluvial) Systems

Though often overlooked because of their flow variations, desert streams are major agents of erosion. They erode, carry, and deposit sediment. Although they carry only about one-millionth of the Earth's water, they are the most important agents of surface change, particularly in Zion National Park. A stream erodes down through uplifted land toward base level, the lowest level to which it can erode its channel. Worldwide base level is sea level; however, streams often encounter temporary base levels, such as lakes, that halt their downward erosion for considerable periods of time.

Stream behavior is a very dynamic geological process. Streams respond immediately to changes in their environment. When a summer cloudburst dumps several inches of rain on a drainage basin, the stream quickly rises, flows rapidly, erodes more sediment, and deposits that sediment downstream. In an unchanging environment a stream's gradient, or slope, is at such an angle that a stream would flow just swiftly enough to transport all the sediment supplied to it by its drainage basin, resulting in little net erosion or deposition of sediment. However, when a stream's gradient is changed, by transgression or regression of the sea, for example, the stream responds appropriately. A lowering of base level causes the stream to cut downward; a rise causes it to deposit coarse sediments onto its bed.

[55]

ABOVE: Water follows a distinct channel formed by a joint in the sandstone in the Left Fork of North Creek.

RIGHT: A V-shaped Zion Canyon appears in the waning light of day as sunlight glistens on the Virgin River.

The principal result of a stream's erosive power is the creation and deepening of valleys. The composition of a valley's rocks and sediments strongly influences the stream's effect. More resistant rocks (like well-cemented sandstone) erode slowly, while loose, less resistant materials are eroded more quickly.

A stream left to its own processes would cut vertically, forming near-vertical walls. However, most river valleys have a distinctive V-shape (see illustrations pg. 66). Why is this so? The answer lies in the relationship between a stream's downcutting and related geologic processes. For example, although downcutting by the Colorado River is largely responsible for the 5,940 feet (1800 m) of vertical relief in the Grand Canyon, landslides and overland flow eroded the 13-mile (21-km) width.

Streams erode their channels by abrasion, hydraulic lifting, and dissolution. Abrasion is the scouring of a streambed by transported sediment in the stream. This is particularly true in swift flowing, sediment-laden floodwaters. Hydraulic lifting, which is erosion under water pressure, occurs when turbulent stream flow dislodges sediment and loosens large chunks of rock. This process is most active during high-velocity floods. As an example, a 1923 flood along the Wasatch Mountain front of central Utah lifted 90-ton (82-metric ton) boulders and transported them more than five miles (eight km) downstream. When a stream flows across and dissolves soluble rocks, such as limestone, dissolution also contributes to stream erosion. The Niagara River, which flows between Lake Erie and Lake Ontario, carries 60 tons (55 metric tons) of dissolved rock over Niagara Falls every minute.

The processes of erosion by running water are very much in evidence in Zion National Park. After the Colorado Plateau rose from its near sea level position, its western edge broke into large blocks bounded by faults. Over time, movement on these faults caused thousands of feet of displacement and tilted the rocks of the Colorado Plateau gently toward the northeast. The Virgin River and its tributaries removed thousands of feet of strata from the block that contains Zion National Park. The erosion

ABOVE: The view from Lava Point on the Kolob Plateau offers a stunning perspective on how the Virgin River and its tributaries carved Zion Canyon. Zion National Park owes its steep-walled canyons to the interaction between the upward movement of the landscape and the erosional work of the river.

RIGHT: The North Fork of the Virgin River flows year-round. Despite its significant gradient, the Virgin River is a small, gently flowing stream that runs through Zion National Park at an average of 100 cubic feet (30 cm³) per second.

was more rapid near the western boundary of this block, which created the current network of west-draining canyons.

Zion National Park, with its spectacular steep-walled canyons and unlimited vistas, owes its very existence to the interaction between the upward movement of these blocks and the erosional work of the Virgin River. Zion Canyon was carved by the Virgin River.

The Virgin River originates on Cedar Mountain in southwestern Utah, flows through the northwestern corner of Arizona and into Nevada where it joins the Colorado River at Lake Mead. The Virgin River has a much steeper gradient than the Colorado, which has contributed significantly to its erosive power. It is comprised of two main branches: the North Fork (Mukuntuweap Canyon), which originates from springs located outside the park's north boundary, and the East Fork (Parunuweap Canyon), which also originates from springs located east of the North Fork. The Virgin River is approximately 160 miles (258 km) long, and empties into Lake Mead. The total loss in elevation along its course is about 7,800 feet (2,340 m), an average gradient of approximately 48 feet per mile (9 m per km). The gradient in Zion Canyon proper is about 71 feet per mile (13 m per km).

Despite its significant gradient, the Virgin River is a small, gently flowing stream. It is not difficult to wade across during most of the year and it flows through the park at an average of 100 cubic feet (30 m³) per second (cfs). It is hard to imagine that this stream could have eroded such an immense canyon as Zion. Although it appears to be clear much of the year the Virgin River transports an estimated one million tons, or more, of rock waste each year. The flow in Zion Canyon is quite variable and the 65-year record suggests a peak flow range from 20 to 9,150 cfs (0.6 to 256 m³ per sec). During the high

ABOVE: A group of canyoneers at the foot of a sheer wall in the Right Fork of North Creek.

LEFT: The April 1995 slide in the lower portion of Sand Bench caused the Virgin River to wash out 600 feet (180 m) of the Zion Canyon Scenic Drive.

flow periods the amount of material transported is staggering. Normal flow carries approximately 120 cubic yards (29 m³) of suspended sediment each 24-hour period, 43,800 cubic yards (10,585 m³) per year. It's estimated that a flood of ten times normal flow carries two thousand times more rock waste. That means one flood event can result in more sediment removal than an entire year at normal flow. When the wetter climates of the ice age and post-ice age periods are considered, the ability of the Virgin River to carve Zion Canyon in so short a time seems far more plausible.

Groundwater Systems

Some precipitation soaks into the ground and slowly migrates down through soil and loose sediment into the underlying rocks. Water that infiltrates the Earth's surface and resides in unconsolidated, or solid rock, is called ground-water. As it percolates downward, this groundwater may find permeable rock layers called aquifers, or it may encounter tightly sealed rocks called aquitards.

Groundwater accumulates below the surface creating a horizon below which all pore space in loose sediment or rock is completely saturated with water. This surface, which fluctuates with climatic conditions, vegetation, and topography, is called the water table. In desert climates like Zion National Park, groundwater resources are the primary agent sustaining living systems.

Groundwater is currently being mined all over the Southwest to supply the drinking water needs of a growing population.

The Navajo Sandstone in Zion National Park is an excellent aquifer that allows infiltrating water to move between sand grains through solid rock. The water typically filters downward until it encounters an aquitard—either a more completely cemented horizon in the Navajo, or the contact between the Navajo and the Kayenta Formations. With its downward progress halted, the groundwater moves horizontally. When its movement intercepts a sheer cliff face, or when erosion lowers the ground surface until the water table is intercepted, a spring results. Springs are very important features in Zion Canyon.

Recharged annually by snowfall in the park's high elevations, springs feed the Virgin River through the dry summer months. Springs are also critical to the existence of numerous micro-climates and environments in park canyons. Take a walk along the Weeping Rock or Gateway to the Narrows Trails and look at the fantastic array of living things associated with these groundwater sources.

In an environment as hostile as Zion Canyon, water is a critical factor, not only as a geologic agent, but to sustain living systems. In Zion National Park, as much as anywhere, one merely has to be observant to recognize that two major hydrologic subsystems—rivers and groundwater—are intimately linked and interdependent. They combined their forces throughout the history of the canyon's development.

ABOVE: Water cascades down the steep wall of Mystery Canyon feeding into the Zion Narrows.

LEFT: A natural spring and grotto near the Virgin River.

RIGHT: Golden columbine (*Aquilegia chrysantha*) forms a hanging garden along a wall in the Zion Narrows.

[63]

"In the desert, the two primary elements are stone and water. Stone comes in abundance, exposed by weathering and a lack of vegetation. It is a canvas. Water crosses this stone with such rarity and ferocity that it tells all of its secrets in the shapes left behind."

—CRAIG CHILDS,
The Desert Cries,
2002

The Zion Narrows are as little as 16 feet (5 m) wide at the bottom and tower as high as a thousand feet (305 m). Here a hiker contemplates the sculpturing of Orderville Canyon.

Canyon

Formation and Evolution

THE TIMING AND MECHANISM for the uplift of the Colorado Plateau, resulting in the formation of Zion Canyon, is somewhat controversial. However, an indisputable fact remains that the Colorado Plateau behaved as a stable, coherent crust segment during most of the last 500 million years. The site of marine sediment deposition during Cretaceous time, it now stands over one mile (two km) above sea level.

The greatest amount of uplift has occurred along the southwestern edge of the Plateau during the last 66 million years. Recent studies suggest that the rate of uplift has varied during this period. Uplift between 25 and 5 million years ago was roughly 130 feet (40 m) per million years.

A hiker negotiates a boulder in Orderville Canyon.

CANYON FORMATION
THREE CANYON PROFILES

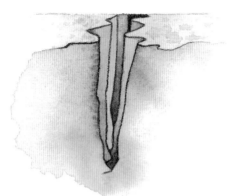

Early stage: Downcutting creates deep,
narrow canyon due to uplift

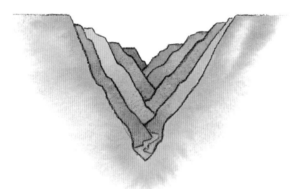

Middle stage: Uplift is moderating. Mass wasting
shapes canyon into characteristic "V"

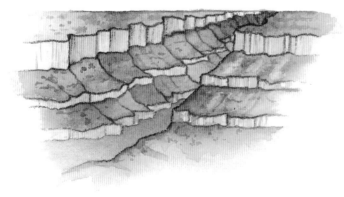

Late stage: Mass wasting continues
until maximum canyon width is achieved

ABOVE: The process of downcutting is particularly evident in the slot canyons of The Narrows at the head of Zion Canyon.

RIGHT: The undulating walls of Orderville Canyon are the result of millions of years of water downcutting through the Navajo Sandstone.

[66]

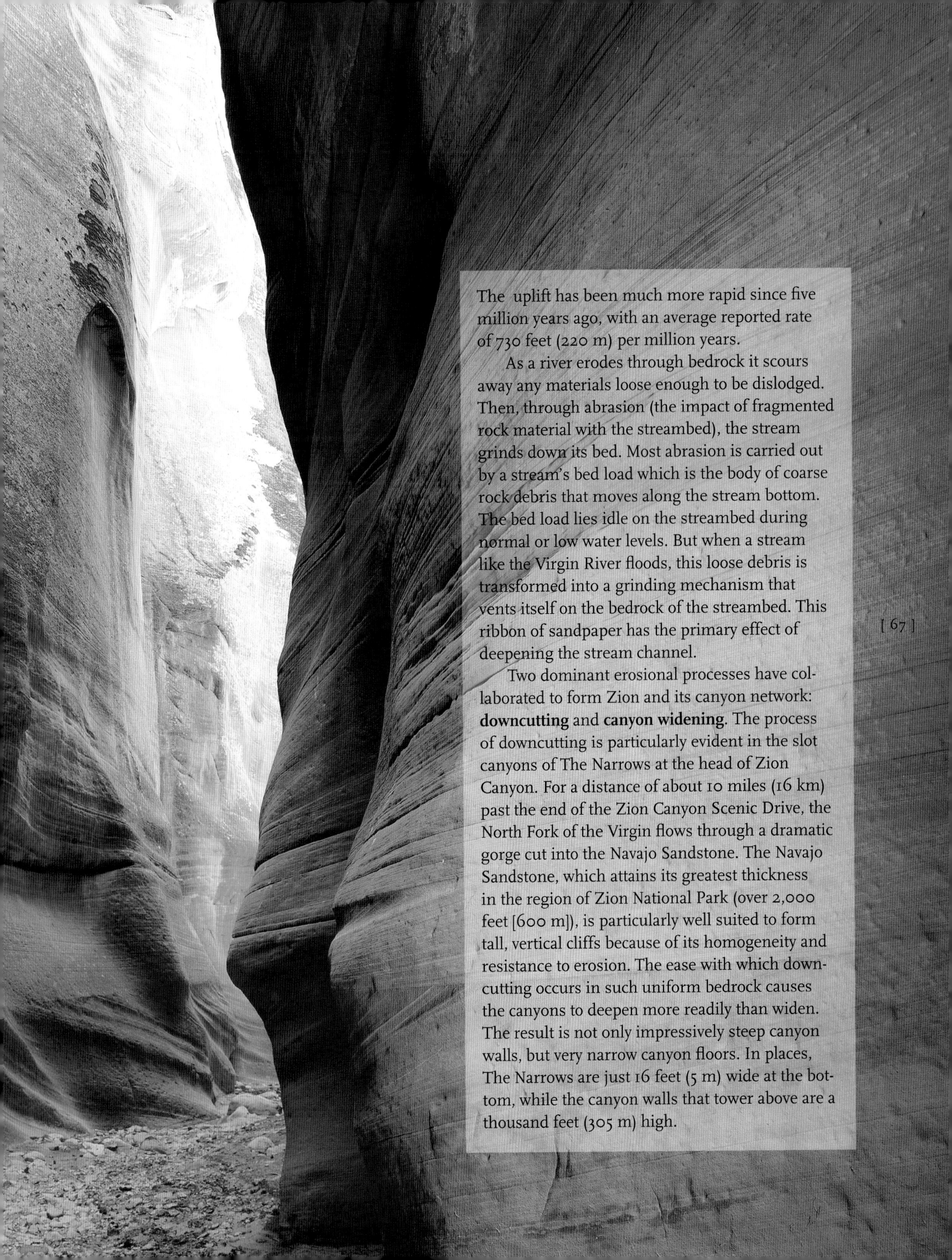

The uplift has been much more rapid since five million years ago, with an average reported rate of 730 feet (220 m) per million years.

As a river erodes through bedrock it scours away any materials loose enough to be dislodged. Then, through abrasion (the impact of fragmented rock material with the streambed), the stream grinds down its bed. Most abrasion is carried out by a stream's bed load which is the body of coarse rock debris that moves along the stream bottom. The bed load lies idle on the streambed during normal or low water levels. But when a stream like the Virgin River floods, this loose debris is transformed into a grinding mechanism that vents itself on the bedrock of the streambed. This ribbon of sandpaper has the primary effect of deepening the stream channel.

Two dominant erosional processes have collaborated to form Zion and its canyon network: **downcutting** and **canyon widening**. The process of downcutting is particularly evident in the slot canyons of The Narrows at the head of Zion Canyon. For a distance of about 10 miles (16 km) past the end of the Zion Canyon Scenic Drive, the North Fork of the Virgin flows through a dramatic gorge cut into the Navajo Sandstone. The Navajo Sandstone, which attains its greatest thickness in the region of Zion National Park (over 2,000 feet [600 m]), is particularly well suited to form tall, vertical cliffs because of its homogeneity and resistance to erosion. The ease with which downcutting occurs in such uniform bedrock causes the canyons to deepen more readily than widen. The result is not only impressively steep canyon walls, but very narrow canyon floors. In places, The Narrows are just 16 feet (5 m) wide at the bottom, while the canyon walls that tower above are a thousand feet (305 m) high.

[67]

An example of erosion in the Left Fork of North Creek, better known as the Subway.

Once the Virgin River cuts deeply enough to encounter the Kayenta Formation, immediately below the Navajo Sandstone, canyon widening begins. Several processes are responsible for canyon widening including rockfall, rock slides, mudflow, and slump. The relatively soft and thin-bedded siltstone, sandstone, and mudstones of the Kayenta Formation are more easily eroded than the overlying Navajo Sandstone. The presence of seeps at the contact of the permeable Navajo and impermeable Kayenta strata also serve to facilitate the undermining and collapse of overlying Navajo strata. As the susceptible Kayenta Formation is removed, the great cliffs of the Navajo Sandstone are undercut and eventually break away, creating massive rock falls on the canyon floor. The many parallel, vertical joints are the factor that controls vertical faces of Navajo Sandstone in

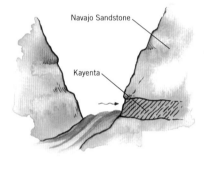

CANYON WIDENING

Canyon widening begins once the Virgin River cuts deeply enough to encounter the Kayenta Formation immediately below the Navajo Sandstone. The relatively soft siltstone and mudstones of the Kayenta Formation are more easily eroded than the overlying Navajo Sandstone. The great cliffs of Navajo Sandstone are undercut and eventually break away.

the canyon's wider portions. These internal fractures are planes of weakness along which rock can break and fall away. Thus, as the canyon widens, the vertical aspect is maintained in the Navajo Sandstone while the valley floor attains greater and greater widths in the more easily eroded underlying strata.

A question often asked by visitors is, "How long did it take to form Zion Canyon?" Recent investigations have unearthed some evidence in the rock record. At a high mesa near Virgin, Utah, a one million-year-old basalt flow caps the mesa at an elevation of 1,300 feet (396 m) above the Virgin River. Thus, it can be deduced that in the vicinity of Virgin, the river has cut down 1,300 feet (396 m) in the past one million years, an average rate of erosion of 1.3 feet (40 cm) per thousand years. By projecting the ancestral Virgin River upstream, it has been suggested that one million years ago in the vicinity of

[68]

CANYON FORMATION

The presence of groundwater—springs and seeps—at the contact of the permeable Navajo Sandstone and the impermeable Kayenta strata, serves to facilitate the undermining and collapse of the Navajo Sandstone. The many parallel, vertical joints in the Navajo Sandstone are the planes of weakness along which the rock can break and fall away. So as the canyon widens, the vertical aspect of the Navajo Sandstone is maintained.

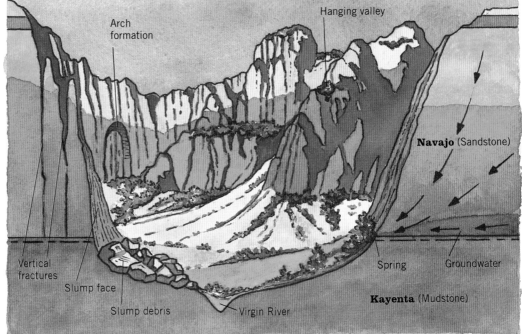

Arch formation

Hanging valley

Navajo (Sandstone)

Vertical fractures

Slump face

Slump debris

Virgin River

Spring

Groundwater

Kayenta (Mudstone)

CANYON FORMATION AND EVOLUTION

A basalt capped ridge above the town of Virgin, just west of the park boundary, was once the bottom of a streambed down which molten lava ran.

INSET: Inverted valleys in and around Zion National Park were formed when molten lava hardened in the bottom of streambeds, forming resistant ridges around which the surrounding landscape eroded.

INVERTED VALLEYS

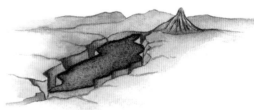

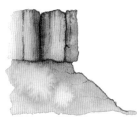

The massive vertical-wall faces seen in Zion Canyon formed well away from the erosive forces of moving water. The vertical joints in the Navajo Sandstone make this possible. These internal fractures are planes of weakness along which the rock, when undercut, can break and fall away, leaving sheer walls of the kind for which Zion is famous.

Zion Lodge, Zion Canyon was only about half as deep as it is today. The Narrows, as we know them, would not have yet been formed, although similar slot canyons probably existed downstream of their current position. Using the assumption of constant downcutting rates over time, the upper half of Zion Canyon was carved between one and two million years ago. Thus, approximately two million years is required for the formation of Zion Canyon below the level of the Carmel Formation. This logic also suggests that a canyon similar to modern Zion Canyon existed in the area of Virgin and Rockville about two million years ago.

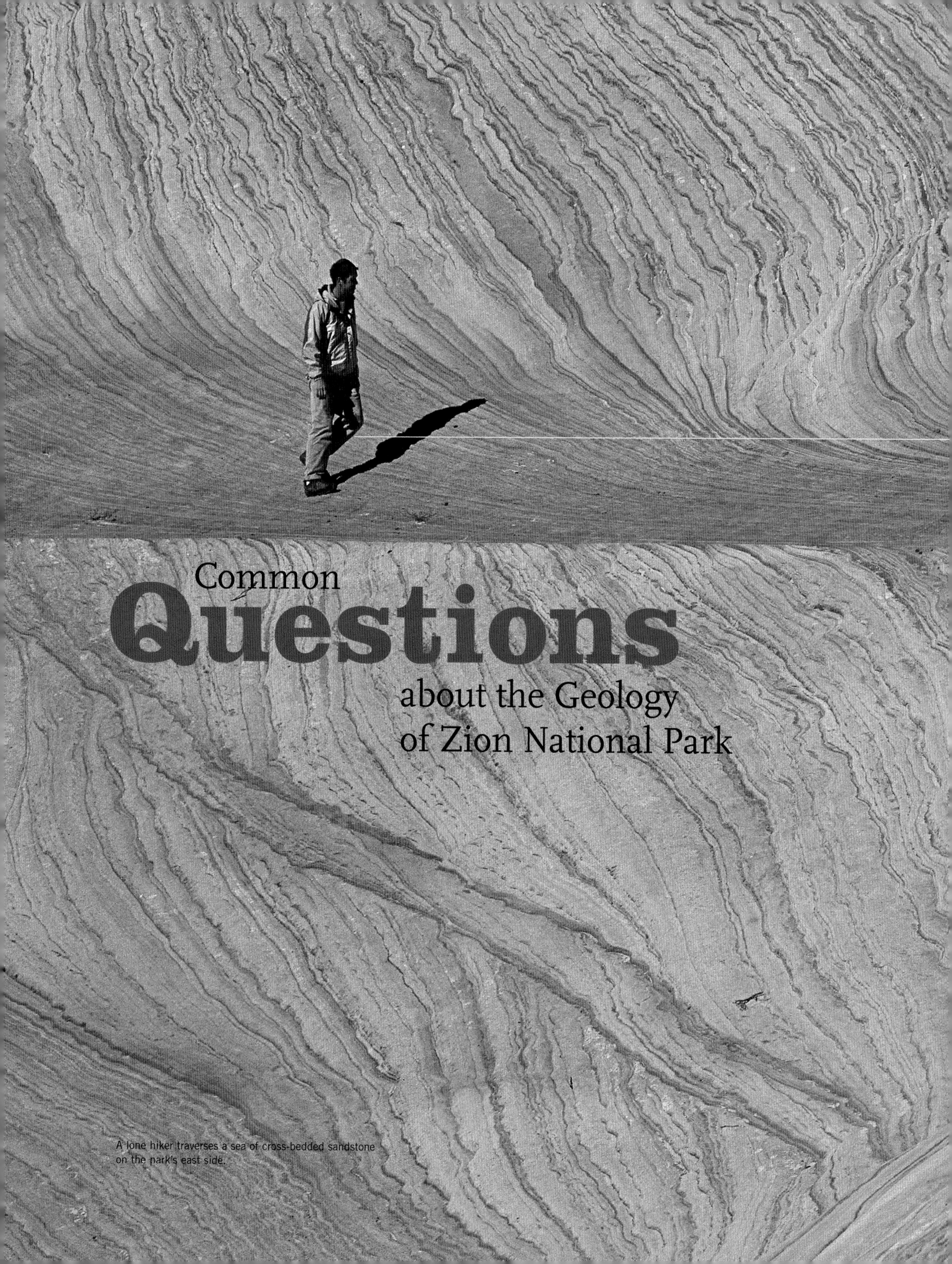

Common
Questions
about the Geology of Zion National Park

A lone hiker traverses a sea of cross-bedded sandstone on the park's east side.

What is cross-bedding?

Normal sedimentary bedding is an arrangement of sediment particles parallel to one another and the surface upon which the sediment is accumulating. A clear break, or bedding plane, is usually visible between adjacent beds. Bedding planes mark the end of one depositional event and the beginning of the next. These changes can be the result of differences in sediment size or transport energy. For example, a river's sediment is a combination of the river's sediment-carrying capability at various times. Interspersed between beds or medium-sized particles, reflecting a stream's normal sediment load, one might find coarse-grained particles deposited during high-energy flooding episodes.

In contrast with normal bedding, which can be prominently seen in the Moenkopi and Chinle Formations, the Navajo Sandstone and upper Kayenta Formations are dominated by cross-beds. Cross-beds are sedimentary layers deposited at an angle to the underlying set of beds.

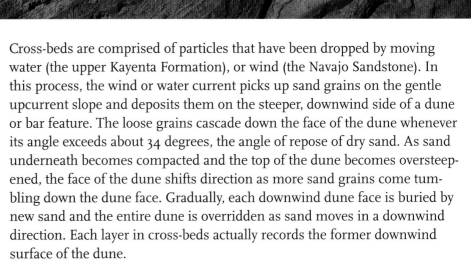

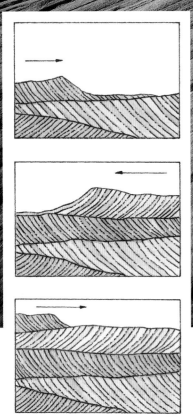

INSET: Cross-beds are sedimentary layers deposited at an angle to the underlying set of beds. In this process, the wind or water has picked up sand grains on the gentle upcurrent slope and deposited them on the steeper downwind side of a dune or bar. Nearly 200 million years later we see the result frozen in the amazing cross-bed patterns in the Navajo Sandstone of Zion National Park.

Cross-beds are comprised of particles that have been dropped by moving water (the upper Kayenta Formation), or wind (the Navajo Sandstone). In this process, the wind or water current picks up sand grains on the gentle upcurrent slope and deposits them on the steeper, downwind side of a dune or bar feature. The loose grains cascade down the face of the dune whenever its angle exceeds about 34 degrees, the angle of repose of dry sand. As sand underneath becomes compacted and the top of the dune becomes oversteepened, the face of the dune shifts direction as more sand grains come tumbling down the dune face. Gradually, each downwind dune face is buried by new sand and the entire dune is overridden as sand moves in a downwind direction. Each layer in cross-beds actually records the former downwind surface of the dune.

Since cross-beds always slope toward the down-current direction, they record the flow direction of the current that deposited them. Cross-beds in the Navajo Sandstone tell us from what direction the prevailing winds of the

OPPOSITE ABOVE: Sandstone patterns on the east side of Zion National Park.

THIS PAGE: Cross-beds in the Navajo Sandstone tell us from what direction the prevailing winds of the Early Jurassic were blowing. Here a hiker observes the fascinating patterns in the Navajo Sandstone.

[75]

Early Jurassic were blowing. The orientation of cross-beds may also help determine paleo wind or water directions in rocks that have been overturned or had their position altered by tectonic forces.

The transition between the Kayenta (river deposited) and Navajo Sandstone (wind deposited) Formations is gradual. As a result, the lower portion of the Navajo Sandstone shows features that suggest a running water influence. In this part of the Navajo the bedding is more parallel than in the upper portion of the formation. There are also places where slumped cross-beds are present. These slump features suggest that at least during deposition of the middle part of the Navajo, the dunes were subject to periodic inundation, perhaps by the nearby sea. As the dune faces became saturated with water, portions became unstable and flowed downward onto the dune face. We see similar slumped cross-beds forming in modern shoreline dune fields.

COMMON QUESTIONS

What causes the color and streaking in Zion Canyon?

One of the most striking aspects of the rocks in Zion National Park is their highly colored appearance. Not only are the rocks themselves remarkably tinted in yellows, reds, whites, and purples, but many of the exposed canyon walls have streaks of color from red to purple-black. Where do these colors come from?

Consider the primary component of the canyon walls: the Navajo Sandstone. Navajo Sandstone is dominantly composed of white-quartz sand grains cemented together with calcium carbonate, silica, and red iron oxide (hematite). The relative variation in the type and amount of cement causes distinct changes in the color of the Navajo Sandstone which has locally been divided into three informal subunits: lower brown, middle pink, and upper white. The boundaries between these subunits are irregular; they rise and fall across cross-bed sets and are difficult to recognize, except from a distance. The lower brown subunit has the strongest cementation and contains more iron oxide than the other subunits. These characteristics cause this portion of the

LEFT: A classic example of a hanging valley with a streak of desert varnish below in Kolob Canyon. The lower brown subunit of the Navajo Sandstone has the strongest cementation and contains more iron oxide than the other subunits. These characteristics cause this portion of the Navajo to weather into sheer cliffs.

BELOW: The upper white subunit of the Navajo Sandstone is exemplified in the Great White Throne. This subunit gets its color from a hydrated and reduced iron oxide called limonite.

OPPOSITE FAR LEFT: A Navajo Sandstone wall in the Kolob Canyons is streaked with desert varnish, which is actually an iron/ manganese oxide coating that remains after water has evaporated.

OPPOSITE LEFT: The Navajo Sandstone is dominantly composed of white-quartz sand grains cemented together with calcium carbonate, silica, and red iron oxide known as hematite.

Navajo to weather into sheer cliffs and hanging valleys, and their associated pour-offs tend to form at the top of this subunit. The middle pink subunit is more uniformly colored by iron oxides and may reflect the unaltered color of the Navajo. The upper white subunit forms the highest cliffs of the Navajo Sandstone in Zion National Park and is exemplified by the Great White Throne. The white color is due to a difference in the chemical character of the iron-rich cement in the sandstone. The middle pink subunit owes its color to the original iron oxide (hematite) cement. The upper white subunit gets its color from a hydrated and reduced iron oxide (limonite).

A question that has intrigued geologists for a long time is: What is the origin of this color variation? Several possible answers have been proposed. One suggestion is that during the accumulation of the sand that would eventually become the Navajo Sandstone, the iron input decreased over time, resulting in a very red lower portion and gradually less red moving toward the top of the unit. Another possible explanation is that the Navajo was originally a

uniform pink, but some process, such as groundwater interaction with the iron minerals, removed the iron color from the upper portion of the formation and concentrated it in the lower portion, producing the distinctive upper white subunit. A much newer theory proposed by geologists at the University of Utah suggests that the Navajo Sandstone was originally uniform in color and contained one of the largest reservoirs of hydrocarbons (natural gas, petroleum, etc.) ever known. During tectonic uplift of the Colorado Plateau at the end of the Mesozoic Era the hydrocarbons escaped, bleaching the color from the upper portion of the Navajo.

There is currently not enough evidence to completely favor one of the above theories to the exclusion of all others. In geology one must often consider multiple possibilities when reconstructing histories and events based on very limited circumstantial evidence.

Besides the imbedded color within the rock, many surfaces on the canyon walls have irregular streaks and coatings. One such coating is desert varnish, which is a generic term for the shiny dark purple to black coating seen on the walls of Zion Canyon. The composition of desert varnish is generally iron/manganese oxide, but its ultimate origin is somewhat controversial. Some geologists suggest that it forms by inorganic processes. Rock in the canyon walls is wetted by rain and water flowing across its surface. As these wetted faces dry, the water evaporates leaving any dissolved chemicals (usually as oxides) in thin coatings. Over time, the layers build up, producing a pronounced coating. Other geologists suggest that the iron/manganese oxide coatings are actually fixed on the rock surface by bacterial growth and metabolism. Over many thousands of years and through the action of millions of such bacteria, a dark, shiny coating builds up on exposed rock surfaces. Streaking can also occur below spring lines and seeps, and usually results from algal growth and the accumulation of white evaporite minerals left behind as mineral-laden spring waters evaporate.

The red streaks which make the Navajo mesa tops appear to drip blood (for example the Altar of Sacrifice), primarily consist of hematite coatings on the surface of the Navajo. These coatings are derived from the overlying red Sinawava Member of the Temple Cap Formation. They form as rain washes this easily eroded blood-red material over the face of the upper Navajo.

Finally, large areas of the canyon walls may be coated with lichens and moss. During the hot summer months these organisms lie dormant, lending subtle shading to the rock surfaces. But during the monsoons the dull lichens burst into brilliant colors and the mosses are transformed to rich greens and browns. Viewed at the right time of year these humble organisms, clinging tenaciously to the rock surfaces, lend an uncommon beauty to the visible canyon ecosystem.

The middle pink subunit of the Navajo Sandstone, predominant in the Kolob Canyons section, is more uniformly colored by iron oxides and may reflect the unaltered color of the formation.

INSETS: Large areas of the canyon walls are coated with lichens and moss. These organisms lie dormant during the hot summer months then burst into brilliant colors during monsoons.

OPPOSITE RIGHT: This massive Navajo Sandstone wall in Taylor Creek is artfully streaked with desert varnish.

? What is weathering?

The first step in erosion is the physical and chemical breakdown of rocks, known as weathering. Weathering has created some fantastic features in Zion National Park, from the popular Checkerboard Mesa to countless unnamed honeycomb and biscuit formations. Rock exposed to the elements cracks, crumbles, partially dissolves, and becomes dirt (more properly called regolith—the unconsolidated material that covers almost all of the Earth's surface and is composed of soil, sediment, and fragments of the bedrock beneath them). Physical weathering is restricted to mechanical processes that break down rock, taking advantage of climate and pre-existing fractures in the rock (called joints). As the processes of physical weathering relentlessly continue over centuries, more and more surface area is exposed to their attack. The interaction of these new exposures with water provides an avenue for chemical attack, and working in concert with mechanical processes, chemical weathering effects rock.

In arid regions the incomplete dissolution of chemicals in rock and the evaporation of mineral-rich groundwater often leaves a coating of salt in fractures and on rock surfaces. As these minerals crystallize within fractures a growth pressure is exerted, resulting in further fragmentation of rock. Such salt coatings, called efflorescences, can be seen along the lower rock exposures of Zion National Park.

When rainwater falls through the atmo-

ABOVE: Hoodoos, such as these in the Kolob Canyons, are capped by a strongly cemented ironstone layer which resists erosion and weathering.

RIGHT: Many rocks in Zion Canyon have a honeycomb look, the result of weathering in the Navajo Sandstone.

As spring water evaporates, mineral deposits, called tufa, are left clinging to the surface of the rock around a spring. Many examples of tufa are found along the trail in the Gateway to the Narrows.

sphere and passes through the soil it readily dissolves carbon dioxide, creating weak carbonic acid. This acid is even more effective at dissolving rock, particularly limestone and calcium carbonate-cemented sandstones and siltstones. The mineral-rich waters move through porous rock like the Navajo Sandstone and, when they encounter less permeable strata, flow to the surface in springs. As the spring water evaporates, mineral deposits, called tufa, are left clinging to the surface around the spring. Excellent examples of tufa development can be found in the Gateway to the Narrows and at Weeping Rock.

How are arches and windows formed?

Like other massive, cross-bedded sandstones, the Navajo Sandstone is highly susceptible to weathering by the process of exfoliation—the peeling off of rock and slabs in a series of concentric layers. The process is accelerated by the presence of regular, parallel, vertical joints in the sandstone. These fractures provide an avenue for weathering agents like water, ice, and frost to attack the rock. At the base of the sandstone, particularly at its contact with an underlying layer, added moisture from seepage evaporates at a slower rate so that the beds at the bottom of an arch weather more rapidly. This undercuts the arch, loosening sheets and blocks of sandstone, and enlarging the overlying arch.

The blind or inset arch above Double Arch Alcove in Kolob Canyons, is the result of weathering by the process of exfoliation, which is the peeling off of rock slabs in a series of concentric layers.

First, the sandstone cliffs overhang because less resistant layers below them have been removed. Then, due to a lack of support, blocks (sometimes several tons) break off along existing fractures and tumble to the canyon floor. Sometimes a blind, or inset arch forms on a cliff face when the upper rock layers remain intact after the lower portion has fallen. An outstanding example of the results of these processes is the Great Arch of Zion, which can be seen from the road as you approach the Zion Tunnel from the west. In an extreme case it is possible for some inset arches to become freestanding, such as Kolob Arch, with a span of 310 feet (95 m), in

the park's Kolob section. This occurs when fractures on both sides of a sandstone block have material removed, leaving a fin (thin blade) of sandstone behind. Normal arch forming processes then cause a window or freestanding arch to form.

The Process Continues

The geologic history of Zion National Park is long and diverse. The accumulation of sedimentary rock over the past several hundred million years has set the stage for tectonic forces, running water, and other canyon modifying processes to continually shape the towers and canyons of Zion. These ongoing, dynamic processes cannot be stemmed and continue to provide evidence of the enormous geological forces at work in the canyons.

What does the future hold for Zion Canyon? Just as the Virgin River has carved its canyons over the past two million years, we can expect the river's processes to continue. We have suggested that Zion Canyon formed in this two-million-year period, and prior to it a similar canyon probably existed where Rockville and Virgin now stand. As long as uplift of the Colorado Plateau continues, the canyons of the Virgin River will continue to migrate upstream. Perhaps in another two million years the current Narrows will look like the southwestern end of Zion Canyon near Springdale, and new narrows will have formed upstream.

Although slow in human time frames, the persistence of geologic processes over the eons of time has demonstrated that continuous change is inevitable. The work of water, wind, and time is ongoing and seemingly endless.

LEFT: In extreme cases it is possible for some inset arches to become freestanding arches, such as Kolob Arch in the Kolob Canyons. With a span of 310 feet (95 m), Kolob Arch is one of the most awesome features in Zion National Park.

BELOW: Just as the Virgin River has slowly but persistently carved Zion Canyon over the past two million years, we can expect the river's processes to continue.

[83]

Geologic
Road Logs

The Temples and Towers of the
Virgin stand 3,800 feet (1,158 m)
above the floor of Zion Canyon.

INSET OPPOSITE: An early visitor
gazes at The Watchman near the
park's south entrance.

Visitor Center to East Park Entrance

[85]

NOTE: *Directions for viewing from the vehicle (or on foot) are given as clock hour, where the front of the vehicle (or the direction you are facing) is always 12:00. A feature directly to the right is then at 3:00, to the left at 9:00. A feature at 1:00 or 2:00 is only slightly to the right of the direction you are facing.*

Mile (km)	Mile (km)	
0.0 (0.0 km)	0.0 (0.0 km)	Visitor Center. Proceed toward State Route 9, the main route through Zion National Park.
0.2 (0.3)	0.2 (0.3)	Junction, Visitor Center, Watchman Campground Drive and State Route 9. Turn right and proceed north on State Route 9 toward the park's east entrance.
0.2 (0.3)	0.4 (0.6)	Entrance to South Campground. The reddish-brown outcrop just above the road to the west is the Dinosaur Canyon Member of the Moenave Formation. The prominent sandstone ledge near the top of the slope is the Springdale Sandstone Member of the Moenave Formation, below which is a mostly covered slope of the Whitmore Point Member of the Moenave. The deep-red slope-former that is poorly exposed above the Springdale Sandstone is the Kayenta Formation. The many large blocks that litter the lower part of this slope are from the Springdale Sandstone. From this point northward, Zion Canyon begins to narrow

The Zion Canyon Visitor Center and shuttle staging area opened in the summer of 2000.

The Zion Human History Museum exhibits hundreds of artifacts relating to the lives of Native American inhabitants and Anglo-American settlers in the Zion Canyon area.

significantly. That is because we have now climbed above the soft, less-resistant, Petrified Forest strata into more resistant layers. Once the river erodes to the level of the Petrified Forest Member it is able to easily erode laterally, causing overlying rocks to slump and widen the canyon downstream.

| 0.3 (0.5) | 0.7 (1.1) |

Oak Creek Bridge.

| 0.2 (0.3) | 0.9 (1.4) |

STOP 1. Oak Creek Canyon. Entrance to Zion Human History Museum. Turn left and enter the parking lot. Park and walk around to the west side of the museum to the patio area and face the sign (you should be looking up Oak Creek Canyon).

On the horizon, located at 12:00, is the Altar of Sacrifice, named for the blood-red streaks staining its surface. This color is derived from the easily eroded Sinawava Member of the Temple Cap Formation which tops the Navajo Sandstone. In the near foreground, at the base of the low hill directly below the Altar of Sacrifice are the remnants of a pioneer irrigation system used to bring water to the abandoned river terraces you see in front of you. These were not only prime farming locations pioneers, but also for the Native Americans who preceded them.

Also on the horizon, at 11:00 is the Sundial, capped by yellow iron-rich sandstone (called ironstone). At 10:00 is a classic view of the West Temple, a sandstone monolith that rises 3,800 feet (1,158 m) above the valley floor, achieving an elevation of 7,810 feet (2,655 m) above sea level.

At 2:00, the Beehives can be seen. These interesting features are in the upper white subunit of the Navajo Sandstone (see the discussion for STOP 2, below) and are capped by a very strongly cemented ironstone layer. The low hill, below the horizon from 9:00–10:00, is composed of units of the Moenave Formation and capped by the Springdale Sandstone Member.

During a paleontologic inventory of Zion National Park conducted in the mid 1990s, students from Southern Utah University discovered some well-preserved dinosaur footprints at the

[87]

head of Oak Creek Canyon in the Dinosaur Canyon Member of the Moenave.

0.6 (1.0)	1.5 (2.4)	Virgin River Bridge.
0.1 (0.2)	1.6 (2.6)	Junction with Zion Canyon Scenic Drive and State Route 9.
0.1 (0.2)	1.7 (2.7)	Unusually large rockfall boulder on right side of road.

0.1 (0.2)   1.8 (2.9)

You are now in Pine Creek Canyon. On the right is a landslide that blocked Pine Creek in the past, forming a pond or small lake. Note the fresh rockfalls off the face of the landslide where it is being undercut by Pine Creek. What you see is important evidence of the ongoing processes of canyon widening. The soft, less-resistant layers are removed by landslide and stream action, eventually undercutting and over-steepening the overlying, more resistant sandstones. The undercutting eventually leaves the sandstone unsupported and it topples from the canyon walls, shearing along existing fractures in the rock. The result is cliff retreat and an ever-widening canyon floor.

The Zion-Mt. Carmel Highway (State Route 9) crosses the Virgin River and heads east up Pine Creek Canyon toward the switchbacks leading to the mile-long Zion-Mt. Carmel Highway Tunnel.

0.1 (0.2)   1.9 (3.0)

The floor of Pine Creek Canyon widens here due to backfilling in the basin upstream of the landslide. This is also one of the few places in the park where the Whitmore Point Member of the Moenave Formation is exposed near a road. The greenish-gray beds about ten feet (three m) thick were deposited in a lake setting, as evidenced by the presence of fossil fish scales. The thicker ledge and overlying rock are the basal part of the Springdale Sandstone Member of the Moenave Formation.

0.2 (0.3)   2.1 (3.4)

Pine Creek Bridge and turnout on the left side of the highway. Beginning here the highway steeply climbs a mostly debris-covered slope over the Moenave and lower Kayenta Formations and then onto a large landslide.

[88]

| 0.5 (0.8) | 2.6 (4.2) | This and several other road cuts over the next 1.5 miles (2.4 km), offer good exposures of the Kayenta Formation. The Kayenta Formation was deposited in a lake and river system that flowed across a broad coastal plain. Most of the dinosaur tracks known in the park are found in the Kayenta Formation. |

| 1.3 (2.1) | 3.9 (6.3) | The Great Arch of Zion, at 10:00. The Great Arch measures 600 feet (283 m) across its base and is 400 feet (122 m) high. Like other massive, cross-bedded sandstones, the Navajo Sandstone is highly susceptible to weathering by the process of exfoliation, the peeling off of rock and slabs in a series of concentric layers, not unlike the layers of an onion. The process is assisted by the presence of regular, parallel, vertical joints in the sandstone. These fractures provide an avenue for weathering agents like water, ice, and frost to attack the rock. At the base of the sandstone, particularly at its contact with an underlying layer, added moisture from seepage evaporates at a slower rate so that the beds at the bottom of an arch weather more rapidly. This undercuts the arch, loosening sheets and blocks of sandstone, and enlarging the overlying arch. |

The Great Arch of Zion is easily visible along Zion-Mt. Carmel Highway (State Route 9) in Pine Creek Canyon. This magnificent inset arch measures 600 feet (283 m) across its base and is 400 feet (122 m) high.

[89]

| 0.3 (0.5) | 4.2 (6.8) | Turnout on right. |

| 0.5 (0.8) | 4.7 (7.6) | **STOP 2.** Switchback and turnout on right. Overview of Zion Canyon.

The Navajo Sandstone can be subdivided into three informal subunits in southern Utah: the lower brown, middle pink, and upper white. These subunits are not true members because they are not defined by stratigraphic boundaries. Rather, they are defined by color changes that result from vertical variations in the amount and type of cement attached to the sand grains of the rock. The boundaries between these subunits are irregular. The brown subunit is the best cemented and contains more iron oxide than the other parts. Hanging valleys tend to form at the top of this subunit. The pink subunit appears more uniformly stained by iron oxides and may repre- |

The historic switchbacks of the Zion-Mt. Carmel Highway (State Route 9) wind their way up the talus of the Kayenta Formation in Pine Creek Canyon to the base of the Navajo Sandstone through which the mile-long tunnel runs. This view is from the end of the Canyon Overlook Trail, directly above the Great Arch of Zion.

sent the unaltered color of the Navajo. The pink subunit is somewhat porous and crumbly, but in places, iron mineralization or cementation is very pronounced and nodules of hard ironstone are scattered on some outcrops.

The white subunit forms the highest cliffs of the Navajo Sandstone in Zion National Park and is best seen in promontories like the Great White Throne. The white color is due to a change in the character of the iron minerals that help to cement the sandstone.

The large bench across the canyon is called Sand Bench. It is actually the collapsed remnant of a large narrow wall, or fin, of Navajo Sandstone that was left behind during erosion of two joints. As canyons cut down along joints on either side of the wall into the weak underlying Kayenta Formation, the wall became increasingly unstable. Finally, about 7,000 years ago, it collapsed, damming Zion Canyon and forming Sentinel Lake. Sentinel Lake was at least 200 feet (61 m) deep and stretched from the Court of the Patriarchs upstream nearly to the Temple of Sinawava. Unlike many Quaternary lakes in Zion National Park, Sentinel Lake was probably full of water year-round.

0.4 (0.6)	5.1 (8.2)	West portal Zion-Mt. Carmel Highway Tunnel. The Zion-Mt. Carmel Highway tunnel is 1.1 miles (1.8 km) long and passes within a few feet of the face of the south wall of Pine Creek Canyon. The tunnel, completed in 1930, was drilled and blasted through the lower 260 feet (79 m) of the Navajo Sandstone. It cost a little over $1,000,000 and was the first million-dollar mile of highway construction in U.S. history. Though modern travelers are not allowed to stop inside the tunnel, there are several galleries and windows in the tunnel that offer views of the adjacent canyon. As you travel through the tunnel the landscape changes from the steep slopes and sheer walls of Zion Canyon to the jointed and sculpted slickrock of the park's east side.

Early visitors enjoy the view from the west portal of the Zion Mt. Carmel Highway Tunnel. The project was completed in 1930.

1.1 (1.8)   6.2 (10.0)

STOP 3. Canyon Overlook Trail. East portal Zion-Mt. Carmel Highway Tunnel and parking area for the Canyon Overlook Trail.

Note the sharp boundary between the brown and pink Navajo Sandstone. This boundary roughly follows a planar surface and represents an ancient oasis (an interdune area) amidst the sand dunes in the Navajo Sandstone. Like a modern oasis, this horizon provides evidence of water and a living community during the time when the Navajo Sandstone was accumulating. This interdune area is exceptionally well exposed along the Canyon Overlook Trail. Take the time to hike the trail, a moderately strenuous hike of 1.0 mile (1.6 km) (if you take this hike, allow approximately one hour). Notice the intense red of the planar oasis surface and look for the evidence of water and living things in the ancient Navajo desert (preserved mudcracks and rootlets). At the overlook you are actually standing above the Great Arch of Zion and looking down Pine Creek Canyon.

Today the tunnel continues to be the only portal on Zion's east side through which millions of visitors enter and leave the park each year.

1.3 (2.1)	7.5 (12.1)	Short tunnel through the Navajo Sandstone.
1.0 (1.6)	8.5 (13.7)	Obvious joints in the Navajo Sandstone at 1:00–3:00.

[91]

RIGHT: The golden leaves of cottonwood trees accent the various sandstone formations in Dry Creek.

0.4 (0.6)	8.9 (14.3)	A blind arch or alcove is forming to the north (8:00). The arch is a natural architectural shape of strength, which is why it is used in so many buildings and bridges. As the Navajo cliffs erode, rockfalls tend to create arches. The self-supporting structure is strong and also dries rapidly after a storm, giving water little time to act on the cement. Arches commonly stand for a long time and may eventually form windows.
0.1 (0.2)	9.0 (14.5)	**STOP 4.** Differential weathering of joints. In several places you will notice large joints along which the level of the sandstone seems to have dropped abruptly on one side. These are good examples of differential weathering. Many people incorrectly believe these are faults since it appears that the sandstone has dropped down on one side. The sandstone is not offset along a fracture as it seems; instead, weathering has acted independently on each side of it. Slight differences in several factors ranging from the dip of the cross-beds to the amount of shade cause differences in erosion rates on opposite sides of the joints. The joint surface may also be coated with iron-manganese oxides which protect it from erosion.
1.3 (2.1)	10.3 (16.6)	Turnout on left.

[93]

LEFT: Dramatic blind arches, such as this massive white arch on Zion National Park's east side, are formed as a result of parallel, vertical joints in the Navajo Sandstone which fracture and provide an avenue for weathering agents like ice and frost to attack the rock.

1.0 (1.6)	11.3 (18.2)	Checkerboard Mesa parking area on left. This park landmark derives its name from the intersecting horizontal and vertical joints that cover its surface. The horizontal joints were produced by differential weathering along original bedding planes. The vertical joints formed locally as the Navajo Sandstone contracted and expanded in response to changes in temperature and moisture.
0.2 (0.3)	11.5 (18.5)	Turnoff to East Rim Trail parking area on left.
0.1 (0.2)	11.6 (18.7)	Zion National Park East Entrance and Fee Station.
0.6 (1.0)	12.2 (19.6)	East Boundary of Zion National Park.

Checkerboard Mesa is the textbook example of horizontal joints produced by differential weathering along original bedding planes, and vertical joints formed locally as the Navajo Sandstone contracted and expanded in response to changes in temperature and moisture.

[94]

BELOW: Shuntavi Butte in the Kolob fingers canyons.

Kolob Canyon Visitor Center to the Scenic Turnout

NOTE: *Directions for viewing from the vehicle (or on foot) are given as clock hour, where the front of the vehicle (or the direction you are facing) is always 12:00. A feature directly to the right is then at 3:00, to the left at 9:00. A feature at 1:00 or 2:00 is only slightly to the right of the direction you are facing.*

Mile (km)	Mile (km)	
0.0 (0.0 km)	0.0 (0.0 km)	**STOP 1.** Exit I-15 at the Kolob Canyons Exit 40 and turn east to the park boundary.

At 12:00, the lower slopes at the base of the Hurricane Cliffs expose fault-sliced fragments of the Moenkopi Formation, evidence of the north-south trending Hurricane Fault zone. The Kolob Canyons road is constructed within the Hurricane Fault zone. Rising above the road, both north and south, is a continuous cliff of Kaibab Limestone. Below it, but difficult to distinguish from it, are the oldest rocks exposed in Zion National Park, the Early Permian Toroweap Formation. Looking south, you can see basalt flows that cap the rocks on the east side of the Hurricane Fault and underlie the juniper-covered plain west of the fault zone. The skyline from 4:00–8:00 is comprised

of intrusive and extrusive exposures of the Middle Tertiary Pine Valley Mountains.

0.1 (0.2)	0.1 (0.2)	The Kolob Canyons Visitor Center.

STOP 2. Parking area, Taylor Creek Trailhead. Stand facing the National Park Service (NPS) information kiosk on the trail.

1.9 (3.1)   2.0 (3.2)

Here (9:00), Taylor Creek has cut through a series of north-south ridges. The ridges are capped by more resistant layers of east-dipping sedimentary rock and are part of the eastern limb of the Kanarra Fold. The westernmost ridge (out of view) is capped by the Virgin Limestone Member of the Moenkopi Formation. The next ridge (10:00) is capped by the Shinarump Conglomerate Member of the Chinle Formation. The third ridge (11:00) is capped by the Springdale Sandstone Member of the Moenave Formation. The ridge above it is also capped by the Springdale Sandstone, evidence of a crustal shortening event that preceded Basin and Range extension and the Hurricane Fault.

Timber Top Mountain and Shuntavi Butte in the Kolob finger canyons. The Navajo Sandstone seems to glow from within at sunset.

At 2:00 is Horse Ranch Mountain, comprised of Navajo Sandstone, Carmel Formation and undifferentiated Cretaceous rock and capped by basalt. Here the Temple Cap Formation is so thin that it is almost absent. The Navajo Sandstone may be more closely observed by hiking to Leda's Cave, 2.0 miles (3.2 km) up Taylor Creek. The cave is located in the lower portion of a stacked double alcove in the Navajo.

[95]

0.9 (1.4)   2.9 (4.7)

At 9:00 is an outcrop of east dipping Springdale Sandstone. Note its characteristics which suggest deposition in a river system. This same unit, located south of here near the town of Leeds, Utah, was the host rock for the Silver Reef mining area.

0.1 (0.2)   3.0 (4.8)

STOP 3. A road cut northwest of the parking area on the right.

This cut through a small hill exposes red and white beds of the Dinosaur Canyon Member of the Moenave Formation. A small fault is visible at 11:00 and offsets a white sandstone bed in the upper Dinosaur Canyon Sandstone which was also deposited in a river, flood-plain environment.

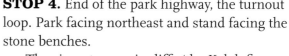

0.2 (0.3)   3.2 (5.1)   At 9:00 is a road cut through exposures of the Kayenta Formation. Note the difference between the appearance of the Kayenta and that of the Dinosaur Canyon Member of the Moenave Formation.

0.4 (0.6)   3.6 (5.8)   Lee Pass Trailhead at 9:00.

1.4 (2.3)   5.0 (8.0)   **STOP 4.** End of the park highway, the turnout loop. Park facing northeast and stand facing the stone benches.

The view at 12:00 is cliffs (the Kolob finger canyons) of Navajo Sandstone complete with cross-beds and extensive jointing. At 1:00, midway up the cliff face, you see hanging valleys related to stream capture of west-flowing streams by south-flowing Timber Creek. At 2:00, a July 1983 rockfall lies beneath the obvious scar it left behind. The vegetated slopes above the Navajo cliffs are on the Carmel Formation. At 4:00 is a great perspective of the Temples and Towers of Zion National Park. Looking further southward, at 4:30, is Smith Mesa with a prominent ridge of Springdale Sandstone just below its top. At 5:00 is Pace Knoll, capped by flat basalt which covers the Moenkopi and Chinle Formations.

[96]

A hiker on Lee Pass Trail in Kolob Canyons.

RIGHT: Sunset in Taylor Creek.

Zion Canyon Shuttle Geologic
Road Guide

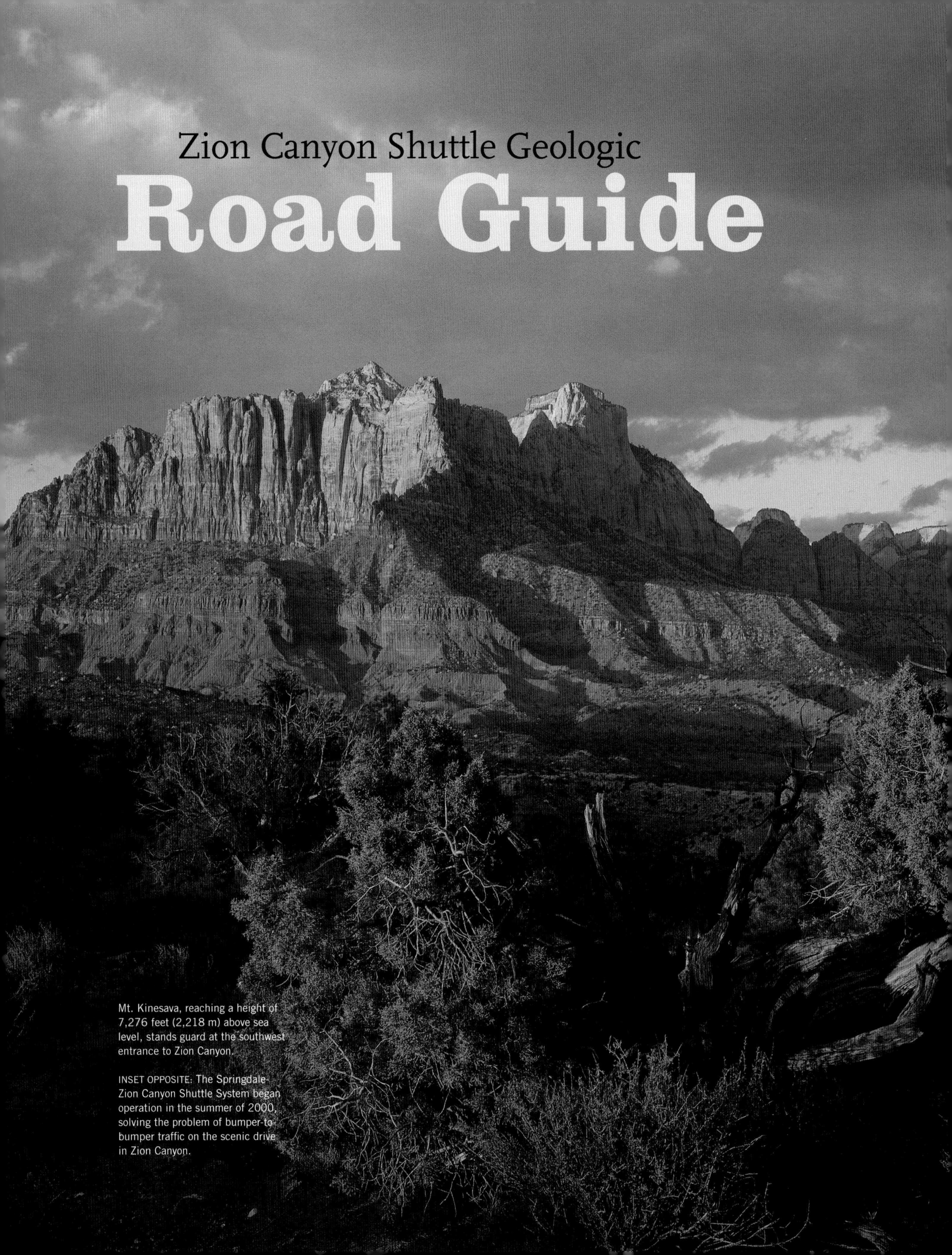

Mt. Kinesava, reaching a height of 7,276 feet (2,218 m) above sea level, stands guard at the southwest entrance to Zion Canyon.

INSET OPPOSITE: The Springdale-Zion Canyon Shuttle System began operation in the summer of 2000, solving the problem of bumper-to-bumper traffic on the scenic drive in Zion Canyon.

Zion Canyon's narrow gorge and towering cliffs attract so many people that the heart of the park can become clogged with traffic. In the summer of 2000, a better way to experience the park was introduced. During much of the year the National Park Service (NPS) prohibits private vehicles on the canyon's six-mile Scenic Drive. Instead, visitors can experience the canyon on the Zion Canyon Shuttle. Only the Scenic Drive is closed to private vehicles; all other roads in the park remain open to private vehicles year-round.

Park in one of the several designated lots in Springdale or at the Zion Canyon Visitor Center, and enjoy a car-free canyon. Buses stop at trailheads, scenic features, Zion Lodge, and the Human History Museum before returning to the Visitor Center. A round-trip on the Zion Canyon Loop is approximately 90 minutes, if you don't get off. Get off and on as often as you like throughout the canyon, hike an inviting trail, take photos, have a picnic, then catch the next bus. To board the Shuttle, follow signs to the main staging area near the Zion Canyon Visitor Center, or get on at any of the designated stops in Springdale or Zion Canyon. The ride is free.

The shuttle system begins on the far southern edge of the town of Springdale at the Majestic View Lodge and Restaurant. During the shuttle season, Springdale is one of the few small towns in America with a free public transportation system.

[100]

NOTE: *Following are notes on the geology visible from several of the stops on the Springdale–Zion Canyon Shuttle. The notes begin with the Springdale Shuttle Stop 9 at the southern end of the town of Springdale and end with notes for the Zion Canyon Shuttle Stop 9 at the north end of the Zion Canyon Scenic Drive. Directions for viewing from the vehicle (or on foot) are given as clock hour, where the front of the vehicle (or the direction you are facing) is always 12:00. A feature directly to the right is then at 3:00, to the left at 9:00. A feature at 1:00 or 2:00 is only slightly to the right of the direction you are facing. Until you enter Zion Canyon at Canyon Junction, all vantage points are located on the east (or right) side of the road.*

Springdale Shuttle: Stop 9

Stand looking northeast (up-canyon) on State Route 9. At 12:00 is Zion Canyon. As you scan the vista in front of you, you see the Towers of the Virgin at 9:30–12:00, which were named by Clarence Dutton of the U.S. Geological Survey.

At 10:00 you see Mt. Kinesava, elevation 7,276 feet (2,218 m), named for a Paiute deity.

At 11:00 is the West Temple, the highest monolith in Zion Canyon. This imposing feature, elevation 7,810 feet (2,380 m), was named by John Wesley Powell, but was called "Mountain without a Trail" by the Paiutes, and Steamboat Mountain by early Mormon settlers.

On the east side of the Virgin River, at 1:00, is the Watchman, elevation 6,555 feet (1,998 m). This rugged peak may have been so named because it stands as a watchman over the canyon.

From 3:00 to 6:30, at the lowest visible level above the Virgin River, you see a prominent exposure of the cliff-forming Shinarump Conglomerate

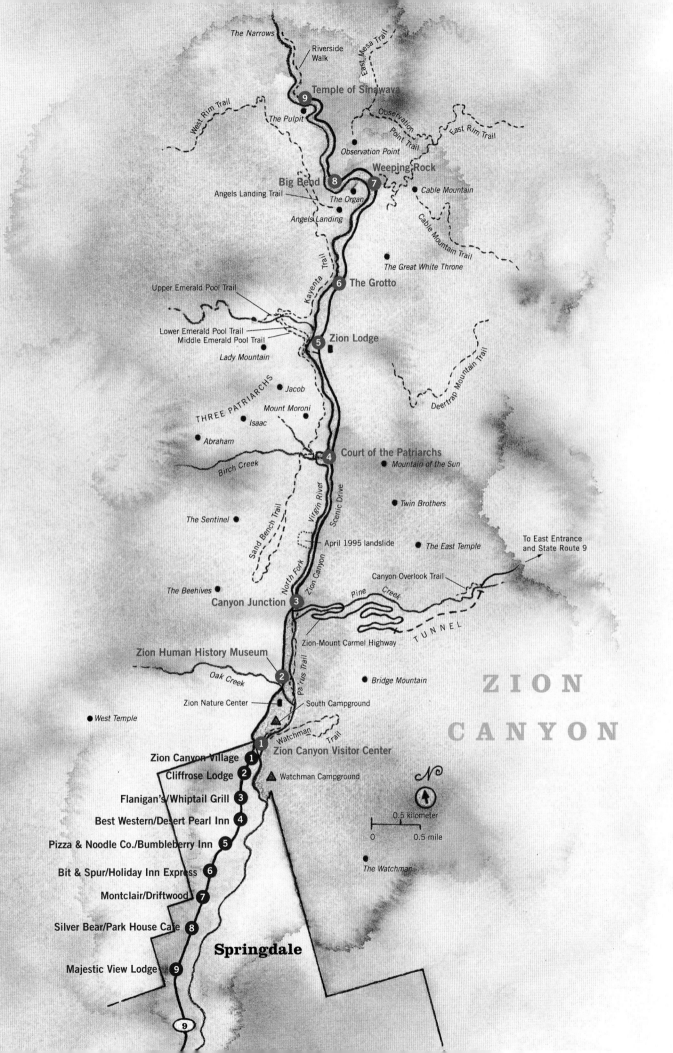

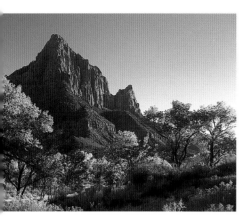

The Watchman, standing sentinel over the south entrance to Zion Canyon, is one of the park's most photographed features.

Member of the Chinle Formation. In fact, the low slopes from 8:00 to 9:30 represent the transition from the overlying Moenave Formation to the underlying Chinle; however, they are covered with eroded and gravity-deposited debris.

Under the imposing heights of Mt. Kinesava, as seen from the town of Springdale, are excellent exposures of the Dinosaur Canyon and Springdale Sandstone Members of the Moenave Formation, as well as the Kayenta and Navajo Sandstone Formations.

Springdale Shuttle: Stop 7

Look northeast toward Springdale, along State Route 9. This represents 12:00. At 9:00, under the imposing heights of Mt. Kinesava, are excellent exposures (from oldest to youngest) of the Dinosaur Canyon and Springdale Sandstone Members of the Moenave Formation, and the Kayenta, and Navajo Sandstone Formations.

At 10:00 you see an excellent view of Mt. Kinesava whose white upper exposures are capped by (oldest to youngest) the Sinawava and White Throne Members of the Temple Cap Formation. This thin, rather localized unit represents a short period of marine conditions at the end of Navajo Sandstone accumulation, followed by an equally brief period of renewed sand deposition. Following the deposition of the Temple Cap Formation, conditions shifted to a marine environment for a long period of time, and the Carmel Formation, most easily seen in the northeastern part of the park, was deposited.

As you look at the exposed Navajo Sandstone that dominates Mt. Kinesava you can see all three informal subunits that make up the Navajo Sandstone—the lower brown, middle pink, and upper white subunits. Also note the extensive fracture system, called joints, developed in the Navajo. This fracture system controls the retreat of the canyon walls and the course that the Virgin River and its tributaries follow.

In the foreground, the low hills are debris-covered slopes of the Dinosaur Canyon Member of the Moenave Formation. The debris that covers them was emplaced by gravity-driven erosional processes collectively called mass wasting (more commonly known as landslides).

From 1:00–3:00, you see the Driftwood Lodge built on an abandoned terrace of the Virgin River.

At 11:00 the Watchman towers above the river and surrounding plain.

Springdale Shuttle: Stop 6

Looking up-canyon, with State Route 9 at 12:00. From 9:00 to 11:00 you see great exposures of Lower Jurassic strata. In particular, note the variegated purple, red, gray, and white slopes of the Whitmore Point and Dinosaur Canyon Members of the Moenave Formation. The Dinosaur Canyon Member represents a thick (nearly 200 feet [60 m]) accumulation of river deposited sediments that incorporates stream channel and flood plain deposits. The Dinosaur Canyon Member is capped by the Springdale Sandstone, also a member of the Moenave Formation.

The Eagle Crags, south of Springdale and just outside Zion National Park, are a remnant of what was once a uniform horizon of Navajo Sandstone.

Looking behind you, at 4:00 is a low ridge of rubble-capped Dinosaur Canyon Member, and at 5:00, you get a great view of the Eagle Crags, a remnant of what was once a uniform horizon of Navajo Sandstone now reduced to a few sharp, isolated, and much eroded peaks.

Zion Canyon is known for its spectacular hanging valleys which tend to form between the middle pink and upper white subunits of the Navajo Sandstone, and are the sites of dramatic waterfalls and pour-offs during storms.

Springdale Shuttle: Stop 5

Looking up-canyon, with State Route 9 at 12:00. The panorama in the cliff face from 1:00–3:00 includes the gradational contact between the underlying Kayenta and overlying Navajo Sandstone Formations. Note the obvious color differences in the Navajo Sandstone. The informal lower brown and middle pink subunits are clearly present.

At 8:30, Mt. Kinesava is clearly visible and at this vantage point you can readily see evidence of an inset arch in formation, and the presence of cross-beds in the sheer canyon walls. The presence of cross-beds records the movement of a large, Sahara-like, sand-sea across this region during Early Jurassic time. The cross-beds themselves represent the former downwind face of the migrating dunes. As you look from the underlying contact with the Kayenta Formation to the top of the Navajo there is an apparent increase in the occurrence of cross-bedding. This observation has been used to support the concept that the climate was evolving during the deposition of the Navajo—from wetter to drier, wind-swept desert conditions.

Also at 8:30 there are several hanging valleys apparent. These valleys, which tend to form between the middle pink and upper white subunits, are the sites of spectacular waterfalls and pour-offs during storms. Zion Canyon takes on an entirely different character during such events, almost not to be believed.

Variegated beds of the lower Moenave Formation (Whitmore Point and Dinosaur Canyon Members) are visible southwest of the Springdale Town Park at 8:00.

Springdale Shuttle: Stop 3

Facing toward the park, the center-line of State Route 9 is 12:00. The magnificent monolith at 12:00 is the West Temple, named by John Wesley Powell during his second expedition to the Colorado Plateau in 1872.

Clearly visible at 11:00 are the rubble and remnants of a magnitude 5.8 earthquake in 1992. The jumbled mound on the west side of the road contains twisted and exposed plumbing. This pile is the toe of a landslide initiated by the earthquake which began advancing on the Cliffrose Lodge (at an initial rate of three or four feet per hour) immediately after the quake, and caused a closure of State Route 9. Above this mound of rubble, in the former Balanced Rock Subdivision, are broken and abandoned homes (one is visible at 10:30). At the base of Mt. Kinesava you can see a visible scar (called an escarpment or scarp by geologists) on the face of the mountain which was created by earthquake ground movement.

As you look east of the highway, the panorama of the east side of Zion National Park begins to open up (from 1:00–4:00).

The most damaging result of the September 2, 1992, earthquake in southwestern Utah was the Balanced Rock Hills landslide in Springdale, just outside the southern border of Zion National Park, which destroyed two water tanks, several storage buildings and three homes, including this one.

Springdale Shuttle: Stop 1

Face the State Route 9 entrance sign to Zion National Park. At 9:00 are rubble-covered slopes of the Dinosaur Canyon Member of the Moenave Formation.

Situated above these low foreground slopes are excellent exposures of the cliff-forming Springdale Sandstone Member of the Moenave Formation. The Springdale Sandstone, named for where it was first studied (Springdale, Utah), was mostly deposited in river channels and contains poorly preserved petrified and carbonized fossil plant materials. The surface of the Springdale Sandstone shows abundant coatings, both the blackish desert varnish (primarily composed of oxides of manganese and iron) and white evaporite coatings (calcium-rich salts crystallized from evaporating runoff and groundwater). Note also the presence of persistent fractures (joints) in the Springdale Sandstone. This fracture system, induced by uplift of the region, is visible to some degree in all exposed canyon walls.

At 3:00 is a bridge that crosses the Virgin River allowing foot traffic to enter the Visitor Center and access the Zion Shuttle System. As you look at the Virgin River, remember that this gently flowing stream is responsible for carving the canyon before you. The river flows through the park at an average of 100 cubic feet (3 m³) per second (cfs). Although it appears to be clear much

The north terminus of the Springdale Shuttle System is on the southern boundary of Zion National Park on the west side of the Virgin River.

of the year, at this flow rate the Virgin River transports an estimated one million tons of rock waste each year. The flow is quite variable and the 65-year record suggests a peak flow range in Zion Canyon from 20 to 9,150 cubic feet (0.6 to 256 m³) per second. During high flow periods the amount of material transported is staggering. Normal flow carries approximately 120 cubic yards (92 m³) of suspended sediment each 24-hour period (43,800 cubic yards [33,500 m³] per year). It has been estimated that a flood of ten times normal flow carries two thousand times more rock waste. That means that one flood event can result in more sediment removal than an entire year at normal flow.

The exposure of Navajo Sandstone visible at 4:00 only contains the lower brown and middle pink subunits. Note the persistence of jointing in the canyon walls.

Zion Canyon Shuttle: Stop 1 Visitor Center

As you stand in front of the shuttle staging area looking up-canyon, the bridge across the Virgin River lies at 12:00. At 2:00 you see Bridge Mountain, elevation 6,814 feet (2,078 m). Note the presence of desert varnish near its base. Bridge Mountain was originally called Crawford Mountain in honor of the William Crawford family, a pioneer family that farmed where the current Human History Museum is located. It is now named for a natural arch—which was called a bridge by pioneers. The arch can be seen from the museum.

At 4:00 you see a small, dry canyon cut under the north face of the Watchman. This canyon was the site of a 1999 debris flow (now referred to as the Watchman Campground debris flow), following a very localized summer cloud burst. Though somewhat overgrown now with vegetation, the rocky debris from this flood event is still visible east of the main shuttle staging area.

Good exposures of the cliff-forming Springdale Sandstone Member of the Moenave Formation are visible at 3:00.

At 11:30 red staining (the result of an iron-oxide mineral called hematite) is visible in the upper white subunit of the Navajo Sandstone. Arch formation is also occurring at 11:30 in the upper white subunit.

From 10:00–8:00 the Dinosaur Canyon and Springdale Sandstone Members of the Moenave Formation and the lower Kayenta Formation are visible. Note the abrupt contact (the plane where one geologic formation ends and another begins) between the Dinosaur Canyon and Springdale Sandstone, and the Springdale Sandstone and the overlying Kayenta Formation.

[107]

The Zion Canyon Visitor Center sits at the foot of The Watchman. Here visitors enjoy interpretive exhibits, the chance to talk with a park ranger, and a variety of books and other interpretive materials available in the Zion Forever Bookstore.

The Zion Human History Museum includes a gallery of exhibits of artifacts depicting human interaction with Zion Canyon.

A number of spectacular Zion Canyon views are available at the Zion Human History Museum. One of the most interesting is a view of the arch on Bridge Mountain.

Zion Canyon Shuttle: Stop 2 Zion Human History Museum

The museum is built on an abandoned river terrace. Such areas have always been important in the human occupation of the canyon. With the museum doors at your back, you are facing 12:00. The first obvious ledge you see is the Springdale Sandstone. The smaller second ledge (11:30) above it is part of the Kayenta Formation. As your eye travels upward try to determine the contact between the upper Kayenta and the overlying Navajo Sandstone Formation. This is difficult to do because the contact is gradual, or gradational—there is no sharp line of distinction.

Located at 12:00 is Bridge Mountain, named for the natural bridge, or arch, located part way up its face (see NPS sign to your left). Like other massive, cross-bedded sandstones, the Navajo Sandstone is highly susceptible to weathering by the process of exfoliation—the peeling off of rock and slabs in a series of concentric layers. These fractures provide an avenue for weathering agents like water, ice, and frost to attack the rock. At the base of the sandstone, particularly at its contact with an underlying layer, added moisture from seepage evaporates at a slower rate so that the beds at the bottom of an arch weather more rapidly. This process undercuts the arch, loosening sheets and blocks of sandstone and enlarges the arch. The sequence starts when the sandstone cliffs begin to overhang, then, due to a lack of support, blocks break off along existing fractures and tumble to the canyon floor. A blind, or inset, arch forms on the cliff face when the upper rock layers remain intact after the lower portion has fallen. In extreme cases inset arches become freestanding, such as Kolob Arch in the Kolob section of the park, with a span of 310 feet (95 m). Freestanding arch formation occurs when fractures on both sides of a sandstone block have material removed leaving a fin (thin blade) of sandstone behind.

The West Temple, the Sundial, and the Alter of Sacrifice are among the Navajo Sandstone pinnacles known as the Temple and Towers of the Virgin, as seen here from near the Zion Human History Museum.

At 10:00 the Temple Cap Formation is visible. It is comprised of the upper, cliff-forming White Throne, and lower, tree-covered, slope-forming Sinawava Members. This formation tops many of the temples and towers of Zion Canyon and is therefore appropriately named.

The Watchman is located at 2:30. Note the vegetation differences from the cacti of the valley floor to the ponderosa pines of the high canyon walls. This variability in vegetation is the result of climatic effects due to increasing elevation (known as relief).

Walk around the outside of the museum to its west side and the patio located there. When you face the existing NPS sign you are facing 12:00, and looking up Oak Creek Canyon.

On the horizon, located at 12:00, is the Altar of Sacrifice (elevation 7,410 feet [2,259 m]), which is named because of the blood-red streaks staining its surface. This color is derived from the easily eroded Sinawava Member of the Temple Cap Formation which tops the Navajo Sandstone. In the near foreground at the base of the low hill directly below the Altar of Sacrifice are the remnants of a pioneer irrigation system that was used to bring water to the abandoned river terraces you see in front of you. These were not only prime farming locations for pioneers, but also for the Native Americans who preceded them.

Also on the horizon, at 11:00, is the Sundial (elevation 7,438 feet [2,267 m]), capped by the upper white subunit of the Navajo and contains a yellow, iron-rich sandstone (called ironstone). It was reportedly so named by the NPS because for many years the people of Grafton regulated their clocks by the early-morning sun shining on its peak.

At 10:00 is a classic view of the West Temple, a sandstone monolith that rises 3,800 feet (1,158 m) above the valley floor, achieving an elevation of 7,810 feet (2,655 m) above sea level.

At 2:00, the Beehives can be seen. These interesting features are composed of the upper white subunit of the Navajo Sandstone and are capped by a very strongly cemented ironstone layer. The low hill below the horizon from 9:00–10:00 is composed of units of the Moenave Formation and capped by the Springdale Sandstone Member.

During a paleontologic inventory of Zion National Park conducted in mid 1990s, students from Southern Utah University discovered some well-preserved dinosaur footprints at the head of Oak Creek Canyon in the Dinosaur Canyon Member of the Moenave.

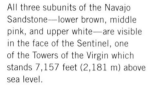

All three subunits of the Navajo Sandstone—lower brown, middle pink, and upper white—are visible in the face of the Sentinel, one of the Towers of the Virgin which stands 7,157 feet (2,181 m) above sea level.

Zion Canyon Shuttle: Stop 3 Canyon Junction

As you look up the road on the east side of the North Fork of the Virgin you are facing 12:00. At 10:00 is another view of the Beehives, elevation 6,825 feet (2,080 m). Note the hanging valley located at the top of the middle pink subunit just south of the Beehives.

At 10:30 you see the Sentinel, the northernmost peak in the group Clarence Dutton called the Towers of the Virgin. At an elevation of 7,157 feet (2,181 m), this imposing promontory appears to be a sentinel guarding Zion Canyon when viewed from up river. All three subunits of the Navajo—lower brown, middle pink, and upper white—are visible in the Sentinel's face. Note the abandoned river terrace at its base which was undoubtedly used for agriculture in the past.

Note the dark staining and desert varnish on the canyon walls at 9:00. Also present are several inset arches and cross-bed sets. Desert varnish is a generic term for the dark purple to black, shiny coating seen on the walls of Zion Canyon. The composition of desert varnish is generally iron/manganese oxide, but its ultimate origin is somewhat controversial. Some geologists suggest that it forms by inorganic

[110]

Desert varnish, made up of iron/manganese oxide, builds up on exposed rock surfaces where water intermittently runs.

processes. The rock in the canyon walls is wetted by rain and water flowing across its surface. As these wetted faces dry, water evaporates leaving any dissolved chemicals (usually as oxides) in thin coatings on the rock surfaces. Over time the layers build up, producing a pronounced coating. Other geologists suggest that the iron/manganese oxide coatings are actually fixed on the rock surface by bacterial growth and metabolism. Over many thousands of years and through the action of millions of such bacteria, a dark, shiny coating builds up on exposed rock surfaces. Streaking can also occur below spring lines and seeps, and usually results from algal growth and the accumulation of white evaporite minerals left behind as mineral-laden spring waters evaporate.

At 3:00 there is evidence of active canyon forming processes in the form of the Pine Creek landslide exposed in the road cut. This mass wasting event occurred in the Kayenta Formation, and the slide continues northward for about 2,000 feet (610 m) along the east side of the road. Landslides like this one are most easily recognized by the precariously balanced boulders and deformed bedding you see in the road cut.

Looking up river, at 11:30, you see the Sand Bench landslide. This feature is not purely a landslide, rather it consists of two parts: (1) a collapsed wall or fin of Navajo Sandstone that blocked Zion Canyon about 7,000 years ago, creating Sentinel Lake, and (2) smaller portions of the old slide mass that have been reactivated (as recently as 1995) and remain active today. Following its most recent slide event, the toe of the slide has been shored up with riprap to help prevent further sliding and damming of the river.

[111]

In April of 1995, the North Fork of the Virgin River undercut the Sentinel Slide triggering a large slide that dammed the river, causing it to cut a new path through the highway.

The three peaks known as the Court of the Patriarchs were named Abraham, Isaac, and Jacob by the Reverend Frederick Vining Fisher, a Methodist minister who passed through Zion Canyon with local guides in 1916.

Zion Canyon Shuttle: Stop 4 Court of the Patriarchs

As you look up the canyon along the road you are facing 12:00. From 8:00–11:00 is the Court of the Patriarchs. These three peaks are named (from the south) Abraham, Isaac, and Jacob (elevations 6,990 feet [2,131 m], 6,825 feet [2,080 m], and 6,831 feet [2,082 m], respectively.) Reverend Frederick Vining Fisher, a Methodist minister passing through Zion Canyon on a day trip in 1916, is credited with naming these peaks. He was accompanied by young Claud Hirschi of Rockville. Birch Creek enters the canyon from the west, just south of Abraham Peak. The abandoned river terrace at this junction is very conspicuous on the west side of the river.

At 3:30 is a hanging valley with a dominant black stain below it marking the trail of seasonal runoff water. Observe the presence of cross-bed sets in the face of this canyon wall.

Outcrops of yellow limonite-enriched zones can be seen at 1:00, at the transition between the middle pink and upper white subunits of the Navajo. This is a fairly persistent phenomenon in the park.

Walk up the trail and view the sandy ledge of the Kayenta Formation below the Navajo Sandstone. You can get a better look at the hanging valley and pour-off viewed previously from this vantage point. Also take time to notice the stones used in the construction of the low wall at the overlook. These stones were probably quarried outside the park. Some have fucoidal markings (horizontal burrows or fossilized feeding traces.) Early investigators described similar markings as impressions of seaweed. The term fucoid was adopted from the generic name of a type of seaweed, but is now used by geologists for any indefinite, trail-, or tunnel-like trace fossils.

Zion Canyon Shuttle: Stop 5 Zion Lodge

Standing with the shuttle stop restrooms immediately behind, you are facing 12:00. Directly in front of you is a panel of Navajo Sandstone showing the weathered character of this unit.

Cross-beds can be seen in relief at 3:00, along with several incipient inset arches. At 1:00 there is a hanging valley, again located at the top of the middle pink Navajo subunit.

Also at 1:00, both light and dark staining are obvious. The river has created a very wide valley in this location because it is the nexus of several side canyons, alcoves and the river, including Heaps and Behunin Canyons coming in from the west and northwest, respectively.

Take the time to explore upper Heaps Canyon and the Emerald Pools area. The trailheads are located across the road from the lodge. No matter how little or much you decide to take on, the hikes are exhilarating and well worth the time invested.

The Zion Lodge, first built in 1925, sits at the foot of towering Navajo Sandstone walls midway up the six-mile Zion Canyon Scenic Drive. The area's first pioneer settler, Isaac Behunin, built his cabin near here in 1862.

Zion Canyon Shuttle: Stop 6 The Grotto

Standing with the shuttle stop directly behind you, you are facing 12:00. This is the jumping off point for some of the most popular hikes into Zion National Park's backcountry. At 11:00 are the West Rim, Angels Landing, and Kayenta Trailheads. The West Rim and Angels Landing Trails proceed across the river and northwestward, up Refrigerator Canyon and Walter's Wiggles (named for former superintendent Walter Ruesch who helped with their engineering in 1924). Along the trail, hikers move through the upper Kayenta Formation, traverse its contact with the Navajo Sandstone, and then move through the two lower Navajo subunits. These two trails separate in the canyon, with the east fork leading to Angels Landing, a 5,790-foot-high (1,765 m) monolith that overlooks Big Bend in the North Fork of the Virgin River. Angels Landing, also named by Reverend Frederick Vining Fisher in 1916, was so named because one of his traveling companions suggested, "Only an angel could land on it."

The Kayenta Trail goes south, across the Kayenta Formation, and takes the hiker to Emerald Pools, a series of three pools created below a small waterfall issuing from a hanging valley at the head of Heaps Canyon.

At 11:00 you see cross-beds in the middle pink subunit of the Navajo. As you look in the direction of 1:30, the West Rim and Angels Landing Trails enter a cut at the mouth of Refrigerator Canyon. The razorback peak located at 2:00 is Angels Landing.

Looking in the 4:00 direction you see the south exposure of the Great White Throne, arguably the poster child of Zion Canyon's temples and towers. At 6:00, fractures in the sandstone walls demonstrate the controls that the existing fracture system exerts on canyon formation.

Another exposure of the Temple Cap Formation is visible at 9:00.

LEFT: The dancing waters of Emerald Pools—a magical destination reachable by a short hike on a well-maintained trail beginning at Zion Lodge.

BELOW: The top of Angels Landing is accessible via the West Rim Trail, starting at The Grotto. The hike is a five-mile round-trip on a trail that is strenuous and steep with narrow sections and unprotected dropoffs.

[115]

Zion Canyon Shuttle: Stop 7 Weeping Rock

With the shuttle stop immediately behind you, facing away from the Weeping Rock parking lot, you are looking at 12:00. This is the jumping off point for the East Rim Trail which takes the visitor to Zion's East Rim, Echo Canyon, Hidden Canyon and Observation Point.

As you look in the 11:00 direction, a different view of Angels Landing is available. From 12:00–2:00, you see the broad meander in the Virgin River known as Big Bend.

Arch formation is occurring in the cliff face at 2:00 and 3:00, and the 2:00 inset arch hosts a granary built by ancient Native American inhabitants, probably to store cultivated grains.

Located at 9:00 is a more classic view of the Great White Throne, elevation 6,744 feet (2,034 m). This feature is one of the most photographed in the park. Named by Reverend Frederick Vining Fisher during his 1916 trip, his journal records that his party, which included a photographer named Ethelbert Bingham and a local muleskinner, Claud Hirschi, had journeyed up river as far as the Temple of Sinawava. As they turned to head back down the canyon, the setting sun highlighted the east canyon walls. Reverend Fisher records: "The clouds were lowering, thunder rolling and a storm brewing and a possible cloudburst and we knew we must go. One of the boys, in trying to take a picture, had fallen into the river—camera and all. And, I fished him out. Then we stood quietly musing and Hirschi said, 'Well, I guess it's the end,' and slowly turned around to go back, when I heard him say, 'Oh, look! What is that? Name that, Doctor!' I swung about and the sight of the Great White Throne, in all its sublime majesty, hit me in the face. I was stunned. For a moment no one spoke. Then I said, 'Well, boys, there is only one name for that. I have been looking for it all my life. I never expected to find it in America or on the Earth itself. That is the Great White Throne!'"

[116]

ABOVE: A dripping spring feeds the hanging garden of ferns and mosses at Weeping Rock, reachable on a half-mile round-trip path beginning at the Weeping Rock shuttle stop. The Great White Throne stands in the background.

RIGHT: The historic cable works, built in 1901. A remnant of the structure is still visible at the top of Cable Mountain above Weeping Rock.

RIGHT: The Great White Throne stands 2,350 feet (716 m) above the Virgin River.

ABOVE: The Zion Canyon shuttle system, instituted in the summer of 2000, has reduced traffic in the canyon and helped improve the visitor experience during the park's busiest season.

Echo Canyon forms a hanging valley at 6:00 where it joins Zion Canyon. Looming above Echo Canyon, at 6:30, you see Cable Mountain, elevation 6,496 feet (1,980 m). This promontory was named for the pioneer cable works (remnants of which can still be seen on the skyline) that brought lumber down from the mesa on a 3,300-foot (1,006-m) cable to near where you are standing. The cable works operated from 1900 into the 1920s. At 7:00, the entrance to Hidden Canyon is visible.

Take the time to walk up the short trail to Weeping Rock. Look for spring lines indicated by the rows of vegetation clinging to the canyon walls. These communities of plants are sometimes referred to as hanging gardens and are commonly associated with springs. Springs form when infiltrating rainwater moves by gravity through porous and permeable sandstone. When the downward progress of water movement is halted by the presence of less porous/permeable layers, the groundwater begins to move horizontally. When its movement intercepts a sheer cliff face, or when erosion lowers the ground surface until the water table is intercepted, a spring results. At Weeping Rock the spring feeds the small stream you cross to walk up the trail. Springs are necessary to insure year-round stream flow. Springs are also factors in arch formation and the canyon-widening process.

The springs at Weeping Rock, as elsewhere in the park, contain significant amounts of dissolved calcium and other minerals. As this mineral-rich groundwater seeps to the surface, the solubility of its mineral content is decreased as carbon dioxide escapes or is removed by plants. As a result, this hard water leaves behind a sponge-like limestone deposit called tufa, recognized by the presence of small holes or pore spaces.

[117]

The Zion Canyon shuttle sweeps around Big Bend near the north end of Zion Canyon.

OPPOSITE RIGHT: The sandstone walls near Big Bend are a climbing paradise. Legendary climber Joe French on Disco Inferno.

BELOW: A room with a view—portaledge bivouacs for climbers on Moonlight Buttress above Big Bend.

Zion Canyon Shuttle: Stop 8 Big Bend

Stand on the downstream side of the shuttle stop with the stop immediately behind you and look downstream on the Virgin River. This is 12:00. At 12:30 is a northern view of the Great White Throne.

From 1:00–4:00 there is a sheer canyon wall topped by a hanging valley. You can see desert varnish and other staining on its surface, cross-bed sets, and intersecting joint sets.

Cable Mountain can be seen again at 11:00, with the ruined cable works clearly outlined in relief against the sky.

Looking in the 8:00 direction, the crumbly, slope-forming portion of the Kayenta Formation is visible. The Kayenta was also deposited in a river system, but there seem to be more flood plain deposits (forming rounded slopes), and fewer channel deposits (sandstone ledges) in the Zion Canyon area. It is this fundamental difference between the Kayenta and Navajo Sandstones that controls canyon formation. In the Big Bend area, the Kayenta, which is highly susceptible to erosion, is at river level. As the river wanders across its flood-plain, particularly during flooding, the Kayenta is removed by erosion, leaving the Navajo Sandstone undercut and over-steepened. This contributes to the instability of the cliff faces, and the existing fracture systems become the logical places for canyon wall failure to occur. Over time, these factors cause the canyon walls to retreat, widening the canyon where the Kayenta Formation is at river level.

Once again, note the gradual transition from the upper Kayenta to the overlying Navajo Sandstone. The canyon walls around you abound with desert varnish, inset arches, cross-beds, and fractures.

The Zion Canyon Shuttle reaches its northern terminus at Temple of Sinawava where the Riverside Walk begins.

Zion Canyon Shuttle: Stop 9 Temple of Sinawava

Stand facing the Riverside Walk sign, facing upstream toward The Narrows. This is 12:00. Directly in front of you is the Gateway to The Narrows. At this point, the level of the river is above the contact between the Navajo Sandstone and the underlying Kayenta Formation. The rock characteristics of the Navajo are relatively uniform, with only the fracture systems they contain providing preferred pathways for erosion. Consequently as the Colorado Plateau rises, it is easier for the river to incise its channel downward than to widen its channel through lateral cutting. The result is the strikingly narrow slot canyons ahead of you.

At 3:00, you can observe yet another hanging valley with staining below its pour-off and incipient inset arch formation occurring.

At 5:00, The Pulpit is visible. The entire rock amphitheater in which you stand is called the Temple of Sinawava. Named by Douglas White, a Union Pacific Railroad publicist in 1913, the name was given to honor Sinawava, another Paiute deity. Looking in the 6:00 direction is the Great White Throne.

A fine example of the canyon-widening processes continuing to work is obvious at 7:00, where large, detached slabs of Navajo Sandstone can be seen at the base of the canyon walls. The notching of the cliff face, just above the tops of the trees at 8:00, represents the scouring of the canyon walls when the river was at a higher level than we find it today. This provides evidence of uplift of the Colorado Plateau and the erosive power of the mighty Virgin River in Zion Canyon.

One of the questions often asked by visitors to the park is "How long did it take to form Zion Canyon?" Recent investigations have revealed some evidence in the rock record. At a high mesa near Virgin, Utah, a one-million-year old basalt flow caps the mesa at an elevation of 1,300 feet (396 m) above

The Pulpit, a spire of Navajo Sandstone standing above the Virgin River, is part of the magnificent rock amphitheater known as the Temple of Sinawava at the north end of the Zion Canyon Scenic Drive.

[121]

the Virgin River. Thus, it can be concluded that in the vicinity of Virgin, the river has cut down 1,300 feet (396 m) in the past one million years, an average rate of erosion of 1.3 feet (40 cm) per thousand years. By projecting the ancestral Virgin River upstream, it has been suggested that one million years ago, in the vicinity of Zion Lodge, Zion Canyon was only about half as deep as it is today. The Narrows, as we know them, had not yet formed, although similar slot canyons probably existed downstream from their current position. Using the assumption of constant downcutting rates over time, the upper half of Zion Canyon was carved between one and two million years ago. Thus, approximately two million years is required for the formation of Zion Canyon below the level of the Carmel Formation. This logic also suggests that a canyon similar to the modern Zion Canyon existed in the area of Virgin and Rockville about two million years ago.

Glossary

Bentonite at the bottom of the Chinle Formation.

abrasion	The erosion of a stream bed by loose sediment transported by a stream.	**bentonite**	A clay formed from the decomposition of volcanic ash. This mineral has a great ability to absorb water and swell accordingly.
alluvial	Of or pertaining to all deposits resulting from the operations of modern streams; including sediments laid down in riverbeds, flood-plains, lakes and fans at the foot of mountain ranges.	**biostratigraphic**	Bodies of strata containing recognizably distinct fossils.
		calcareous	Containing calcium carbonate (calcite).
basalt	The dark, dense, volcanic rock that makes up most ocean plate material. Basalt is common in volcanic flows and cinder cones.	**calcite**	A mineral, calcium carbonate, $CaCO_3$.
		carbonaceous	Pertaining to, or composed largely of, carbon.

Basalt in Zion National Park is the result of relatively recent volcanic activity.

		caldera	A large depression typically caused by the collapse or ejection of the summit area of a volcano.
		cfs	Cubic feet per second. One cubic foot of water flowing past a certain point in one second.
base level	The lowest level to which a stream can erode its channel.	**chert**	A compact, siliceous rock formed from cryptocrystalline varieties of silica (quartz).
bed load	A body of coarse rock debris that moves along a stream channel whenever enough hydraulic energy is available to lift it.		

[122]

Conchoidal fracturing on the north face of Angels Landing.

conchoidal
Refers to a type of rock or mineral fracture that gives a smoothly curved surface. The term derives from the curve of a conch shell.

Concretions known as Moki marbles.

concretion
A nodular or irregular concentration of mineral constituents in sedimentary rocks, developed by localized deposition of material from solution, generally around a central nucleus.

cyclicity
Tendency of some sedimentary processes to repeat over time resulting in repetitive, stacked strata.

deposition
The accumulation of earth materials in topographic low points, like valleys and ocean basins. These deposited materials eventually become sedimentary rock.

dissolution
A stream's removal of soluble materials from rocks over which it flows.

eolian
A term applied to the wind's erosive action, and to deposits which are due to the wind's transporting action.

erosion
The process by which particles of rock and soil are loosened and then transported elsewhere by wind, water, ice, or gravity.

evaporitic
Climatic conditions leading to the formation of evaporites; sedimentary rocks formed by material deposited from solution during the evaporation of water.

exfoliation
The peeling of rock and slabs in a series of concentric layers, much like the interior of an onion, which then break loose and fall from the steep faces of canyon walls.

Sandstone bedding on the east side of Zion National Park.

[123]

felsic
A general term applied to light-colored igneous rocks containing an abundance of feldspar and quartz.

fluvial
A term referring to streams, stream action, and stream deposits.

formation
The basic litho (rock) stratigraphic unit. A unit of strata that is mappable and has distinctive upper and lower boundaries.

Folds formed by compressive rock thrusting. Rock folded in Taylor Creek.

fold

A bend or warping in bedding or foliation in rock, resulting from deformation.

fossil

A remnant, imprint, or trace of an ancient organism preserved in the Earth's crust.

gradient

The vertical drop in a stream's elevation over a given horizontal distance, expressed as an angle.

hydraulic lifting

The erosion of a streambed by water pressure.

igneous

Rock formed from solidification of magma on the Earth's surface or within the crust.

lacustrine

Produced by, or belonging to, lakes.

lithology

The physical characteristics of rocks.

Sunset on Navajo Sandstone.

lithospheric plates

One of the large, thin, rigid units that make up the Earth's outermost layer. These plates may be continental, oceanic, or both.

marine

Of, belonging to, or caused by, the sea.

mass wasting

Loosening and downslope movement of soil and rock in response to gravitational stress; rates of movement range from creep (very slow) to very rapid falls and slides.

member

A subunit of the formation. Multiple members make up a formation.

metamorphic

Rock that has been transformed from preexisting rock into texturally or mineralogically distinct new rock as a result of high temperature, high pressure, or both, but without the rock melting in the process.

orogeny

The process of forming mountains, especially by folding and thrust faulting; an episode of mountain building.

oxidation

The loss of electrons by an element or ion due to chemical interaction (weathering); so named because the elements commonly combine with oxygen.

paleo-

A prefix, meaning ancient.

Pangaea

The name proposed by Alfred Wegener for a supercontinent that existed at the end of the Paleozoic Era and consisted of all of Earth's landmasses. The word pangaea is a Greek term meaning "all the Earth."

A representation of Pangaea.

playa	The flat central area of an undrained desert basin.	**slump**	The downward and outward slide of loose rock debris along a concave slip plane.
reduction	The gain of electrons by an element or ion due to chemical weathering contrasted with oxidation.	**subaerial**	Formed, existing, or taking place on the land surface contrasted with subaqueous.
regolith	The unconsolidated material that covers almost all of Earth's surface and is composed of soil, sediment, and fragments of the bedrock beneath them.	**subduction**	The sinking of an oceanic plate edge as a result of collision with a plate of lesser density. Subduction can cause earthquakes and create volcanic chains.
regression	The opposite of marine transgression. The withdrawal of the sea from a continent or coastal area resulting in the emergence of land as sea level falls or the land rises with respect to sea level.	**topography**	The physical features of a district or region, such as are represented on maps. The land's relief and contour.
		transgression	Refers to the migration of a marine shoreline. The invasion of coastal areas or much of a continent by the sea resulting from a rise in sea level or subsidence of the land.
relief	The difference in elevation between a feature's highest and lowest points.		
resistant	Less susceptible to the forces of weathering and erosion. Resistant strata tend to form cliffs and rock faces.	**unconformity**	An erosion surface that marks lapses in the accumulation of earth materials. Missing rock layers represent gaps in the geologic record.
river terrace	Flat area where a river once occupied space.	**volcanic arc**	Mountains formed in part by igneous activity associated with the subduction of oceanic lithosphere beneath a continent. Modern examples include the Andes and Cascade Mountains.
sabkha	A tidally produced salt flat.		

[125]

Sedimentary rock forms the walls of Zion Canyon.

		weathering	The process by which exposure of rock to atmospheric agents (air or moisture) causes them to break down. This process entails little or no movement.

Weathering results from the exposure of rock to water and air.

sedimentary	A rock formed at or near the Earth's surface, and composed of consolidated solid fragments of rock or organic remains.
shale	A sedimentary rock composed of clay-sized particles. Shale usually originates in relatively still waters.

References

of General Interest

BIEK, R. F., WILLIS, G. C., HYLLAND, M. D., AND DOELLING, H. H. "Geology of Zion National Park, Utah." In *Geology of Utah's Parks and Monuments,* edited by D. A. Sprinkel, T. C. Chidsey, Jr., and P. B. Anderson. Bryce Canyon Natural History Association. Utah Geological Association Publication 28. 2003. p. 106–137.

CRAWFORD, J. L. *Zion National Park: Towers of Stone.* Zion Natural History Association, Springdale, Utah. 1988. 47 p.

DECOURTEN, FRANK. *Dinosaurs of Utah.* University of Utah Press, Salt Lake City. 1998. 300 p.

GREGORY, H. E. *Geology and Geography of the Zion Park Region, Utah and Arizona.* U.S. Geological Survey Professional Paper 220. 1950. 200 p.

GREGORY, H. E. *A Geologic and Geographic Sketch of Zion National Park.* Zion-Bryce Natural History Association, Springdale, Utah. 1956. 36 p.

HAMILTON, W. L. *Geological Map of Zion National Park, Utah.* Zion Natural History Association, Springdale, Utah. 1978 (revised 1987). Scale 1:31, 680.

HAMILTON, W. L. *The Sculpturing of Zion.* Zion Natural History Association, Springdale, Utah. 1995. 132 p.

HARRIS, A. G., AND TUTTLE, ESTHER. *Geology of National Parks,* 4th edition. Kendall/Hunt Publishing Company, Dubuque, Iowa. 1990. 652 p.

HAYDE, FRANK R., AND RACHLIS, DAVID. *Zion: The Story Behind The Scenery.* KC Publications, Las Vegas, Nevada. 2003. 64 p.

HINTZE, L. F. *Geologic History of Utah.* Brigham Young University Geology Studies Special Publication 7. 1993. 202 p.

HOPKINS, RALPH LEE. *Hiking the Southwest's Geology: Four Corners Region.* The Mountaineers Books, Seattle, Washington. 2002. 288 p.

Photography
Credits

Sandy Bell: p. 80 (bottom), p. 120

Joe Braun: p. 9, p. 90

Eric Draper: back cover, pages 4–5 (table of contents), p. 53, p. 54 (top right), p. 63 (left), p. 65 (inset), pages 72–73, p. 75, p. 104, p. 108 (bottom), p. 118 (bottom), p. 119, p. 123 (top left)

John Gnass: p. 15 (inset)

Dave Grimes: p. 47, courtesy of Zion National Park

Lyman Hafen: p. 32 (inset), p. 35 (bottom), p. 45, p. 46 (top), p. 70, p. 81 (top), p. 91 (bottom), p. 117 (left), p. 124 (top left)

John K. Hillers: p. 21 (top) courtesy of Smithsonian Office of Anthropology, Bureau of American Ethnology Collection, negative no. 1636

George H. H. Huey: front cover, p. 1, pages 84–85, p. 116 (top), p. 117 (top right)

Nick Jorgensen: p. 10 (top)

M. H. Levy: p. 59

Donn Lusby: p. 8 (inset), p. 12, p. 27, p. 29, p. 43 (top row left, center, and right; middle row left; bottom row left and center), p. 44 (top), pages 52–53, p. 54 (left), p. 55 (bottom), p. 57, p. 68, p. 83, p. 87 (bottom), p. 88, p. 89, p. 92, p. 93, p. 94 (top and bottom), p. 96, p. 100, p. 102 (left and right), p. 103, p. 109, p. 110, p. 111 (top), p. 112, p. 122 (left), p. 123 (right)

David Pettit: pages 16–17, p. 18, p. 33 (inset), p. 36, p. 42, p. 60, pages 64–65, p. 77 (bottom), p. 79, p. 98–99, p. 114, p. 122 (right), p. 125 (left)

Michael Plyler: p. 10 (bottom), p. 19 (top right and bottom left) both courtesy of Zion National Park Museum [compass: ZION 953, knapsack: ZION 949], p. 20 (top) courtesy of Zion National Park Museum [ZION 266], p. 21 (bottom) courtesy of Zion National Park Museum [ZION 10754], p. 32 (background), p. 34 (bottom), pages 40–41, p. 46 (bottom), p. 50, p. 55 (top), p. 56, p. 58, p. 69, p. 77 (top), p. 78 (insets top and bottom), p. 80 (top), p. 87 (top), p. 99 (inset), p. 106, p. 107, p. 108 (top), p. 113, p. 118 (top), p. 121, p. 123 (bottom left), p. 125 (right)

Tom Till: pages 6–7, p. 11, p. 51, p. 62, p. 63 (bottom), p. 66 (inset), pages 66–67, p. 74, p. 95, p. 97

John Wagner: pages 34–35 (top), p. 71, p. 76 (right), p. 82, p. 115

George Ward: front and back endsheets, pages 2–3 (title pages), pages 14–15, p. 76 (left), p. 78 (background), p. 81 (bottom)

Eric Wunrow: p. 8 (background), p. 13, pages 28–29, p. 31 (left and right), p. 43 (middle row center and right; bottom row right), p. 44 (bottom), p. 54 (bottom right), p. 124 (bottom left)

OTHER PHOTOGRAPHS:

p. 19 courtesy of Bishop Museum, Honolulu, Hawaii

p. 20 courtesy of Zion National Park Museum (ZION 12406); courtesy J. L. Crawford

p. 21 (middle) courtesy of Zion National Park Museum (ZION 12366)

p. 24 (top) courtesy of the Alfred Wegener Institute for Polar and Marine Research, Bremerhaven, Germany

p. 61, p. 85, p. 91 (top), p. 105, p. 111 (bottom) courtesy of Zion National Park

p. 116 (bottom) courtesy of Zion National Park Museum (ZION 14697)

Index

[131]

About the Author

Robert L. Eves is Professor of Geology at Southern Utah University (SUU). He holds a PhD in Geology and Geochemistry from Washington State University and is currently chairman of the Department of Physical Sciences at SUU. A past chairman of the Rocky Mountain Section of the Geological Society of America (GSA), he has presented more than thirty papers at GSA meetings, and is a regular contributor to the *Journal of Geoscience Education*. Having served as president of the Southwestern Section of the National Association of Geology Teachers, Dr. Eves is a dedicated and popular instructor. His love for Zion National Park and its marvelous geology is demonstrated by the fact that he regularly trains park interpreters who share the geologic story of Zion with thousands of visitors every year. He served on the board of directors of Zion Natural History Association for twelve years, with a two-year term as chairman. He and his wife Patricia are the parents of three children. He is committed to his family, his church, and his profession—in that order.

[132]

**ZION NATL PARK
FOREVER PROJECT**

About the Zion National Park Forever Project

The Zion Forever Project traces its roots back to 1929, when a group of citizens living at the gateway to the park formed the Zion Natural History Association and focused their efforts on chronicling the canyon's past and ensuring its future. Reflecting their love and passion for the park they began providing visitors with accurate and inspiring information. Their pamphlets, maps, books, post cards and other interpretive materials were designed to enhance the visitor experience and to educate visitors about the unique flora, fauna, geology, and human history of the canyon. Proceeds generated from sales in the visitor center bookstores went back to produce more and better interpretive products, and fund important projects in the park.

Through the decades the width and breadth of our work has expanded as the park's annual visitor count has grown from tens to hundreds of thousands, and finally, millions. We fund iconic programs in the park and help visitors not only have a more meaningful visit, but truly connect with Zion and return home as life-long advocates.

The love and passion that drove those original citizens to form this organization has recently grown into a much deeper sense of stewardship and responsibility. Our commitment to Zion's future has brought us to a place where all our efforts in publishing, bookstores, fundraising and field programs, are now combined under the banner of the Zion National Park Forever Project. We are here to address and help provide solutions to the park's greatest challenges, and to create a margin of excellence for people like you who love this place so much. Your purchase of this wonderful book on the geology of Zion contributes to this ongoing effort and will make a difference now and forever.

Visit **zionpark.org** to learn how you can continue to make a difference for Zion.